KU-680-893

Leveraged Innovation

Unlocking the Innovation Potential of Strategic Supply

Francis Bidault
Theseus Institute, Rue Albert Einstein, BP 169,
06903 Sophia Antipolis Cedex, France
tel 33 (0) 4 42 92 94 51 00 / bidault@theseus.fr

Charles Despres
Theseus Institute, Rue Albert Einstein, BP 169,
06903 Sophia Antipolis Cedex, France
Ecole Supérieure de Commerce, Marseille-Provence,
Marseille, France
tel 33 (0) 4 42 53 13 19 / despres@escmp.u-3mrs.fr

Christina Butler
London Business School, Sussex Place,
Regent's Park, London NW1 4SA, UK
tel 44 (0)171 262 5050 / c.butler@lbs.ac.uk

First published 1998 by

MACMILLAN PRESS LTD

Houndmills, Basingstoke, Hampshire RG21 6XS

and London

Companies and representatives
throughout the world

ISBN 0–333–74938–3

A catalogue record for this book is available from the British Library.

This book is printed on paper suitable for recycling and
made from fully managed and sustained forest sources.

10 9 8 7 6 5 4 3 2 1

07 06 05 04 03 02 01 00 99 98

Formatted by *The Ascenders Partnership*, Basingstoke

Illustrations by *Ascenders*

Printed in Great Britain by Creative Print & Design (Wales),
Ebbw Vale

Contents

Preface

The great French scientist Pasteur, inventor of vaccination, is reputed to have said, 'Chance only favors the prepared mind.'

Few would debate that luck is a key ingredient of innovation: Penicillin, Post-It notes, Teflon were spawned by fortunate chains of events. Serendipity has been, in fact, an important factor behind many breakthrough innovations. But many of these fortunate and unexpected chains of events, like a mistake in an experiment, would go unnoticed to the unprepared mind. They led to breakthroughs only because prepared minds realized their meaning.

Chance comes in many ways. It may come from within the organization when an innovator or colleague tinkers with his or her research. This is commonly discussed as a form of 'serendipity'. But chance can also be triggered by exposure to 'outsiders'. In this respect, innovation is fostered by an interaction with other companies' research and development. It is now widely accepted that the most innovative firms tend to be the most open to collaboration with other organizations. Companies like Astra Pharmaceuticals of Sweden, owner of the world's best selling drug, Losec, or Hewlett-Packard with its proximity to leading Californian universities are examples of the strong advantage these collaborations provide.

Along with universities, competitors and clients, suppliers are now contributing more as sources of innovation. This has been observed for some time in the automobile industry. The role suppliers can play in product development gained new status due to the competitive edge of Japanese car manufacturers, who extended their reliance on suppliers to levels never before attempted in the West.

In the late 1980s, the practice of involving suppliers in product development was widely emulated by most car makers. The advantage of this approach – now known as Early Supplier Involvement or ESI – soon became 'common sense' but the managerial and organizational implications remained ill-defined for the most part. Many questions lingered in the minds of executives tempted to involve suppliers in their product innovation processes: What are the real benefits of ESI? What are the risks? How can one mobilize a supplier's competencies in product development? How does one deal with the intellectual property arising from a collaboration? How to manage the pricing of parts that are developed with the support of suppliers? How to make sure that managers who are accustomed to bargaining actually embrace a more collaborative mode with suppliers?

The list of issues is long and fascinating because it embraces some of the central issues that companies will face in the 21st Century as they move away from traditional forms of organizations. With the need to develop products that integrate more and more technologies, companies must cross the borders of their traditional industries and join forces with other organizations. Car makers must now offer a product that integrates navigation systems that were once reserved for the military and maritime. Watch makers have to incorporate telecom capabilities in their time-keeping pieces. Vacuum-cleaner manufacturers include features that borrow the know-how of soap makers. The development of product and technology appears to be a boundary-spanning activity *par excellence*.

We realized the importance of these issues during our tenure at IMD, a leading European management institute located in Lausanne, Switzerland. Through our contacts with executives participating in IMD's seminars, we became convinced that the management of 'joint innovation' was at the frontiers of industry. We were fortunate that Professor Thomas Vollmann, who was leading IMD's main research programme at the time, supported our research in this area. The 15 or so blue chip companies that sponsored the programme 'Manufacturing 2000' provided the support for a first research project and agreed to participate along with several others that we were able to study in the three regions of the Triad (Europe, North America and Asia). Our first project aimed at identifying the issues associated with the involvement of suppliers in product innovation.

We undertook a wide survey of the literature in this domain and after identifying the leading authors and organizations, set up a plan to interview them. Meetings took place in Paris, Goteborg, Tokyo and Boston. We had a fascinating meeting with Professor Takahiro Fujimoto of Tokyo University who had co-authored with Professor Kim Clark, now Dean of the Harvard Business School, a seminal book on the differences in product development processes between American, European and Japanese car makers. In another interview, we learned about the implementation of ESI at Renault. Dr Christophe Midler, a senior researcher from Ecole Polytechnique who had conducted a longitudinal study on Renault's development of the Twingo sub-compact car, told us about the challenges encountered by the development team as they extended the engineering responsibilities of suppliers to new lengths.

The car industry, it appeared, was a major source of knowledge in the domain since it had experimented earlier than other industries with the redefinition of the supplier's role. Consulting companies, particularly Andersen Consulting in Paris and Arthur D. Little in Sweden, also contributed substantially to our understanding.

But the best source of learning remained the people who had experienced the ESI challenge. We therefore met with executives who had confronted these issues in their professional life. Our interview list included the leading innovators in the corporate world, such as Bosch and Braun in Germany, Motorola and Xerox in the US and Sony and NEC in Japan. In these companies, we met with executives who had been in charge of development projects ('project leaders') and who had, generally, a substantial experience with ESI … both positive and not so positive. The first contacts also involved colleagues from Purchasing who are normally concerned with all issues pertaining to sourcing. These interviews allowed us to collect a diversity of views about the benefits and drawbacks of integrating suppliers in product innovation. We also discussed the difficulties encountered in their projects when trying to coordinate development activities with their partners.

We did not limit our interviews to 'clients', however. We wanted to get a complete picture by involving 'suppliers' who might have a different perspective from their clients. We conducted research with large technologically advanced companies such as BP Chemicals, Vetrotex (a leading player in glass fibres) and Sumitomo Electric which provides the Japanese car industries with subassemblies. We also included a collection of small plastic moulders such as Seitz through to major international players such as Plastic Omnium and Nypro.

Interviews are valuable in collecting viewpoints and opinions but they are less effective when dealing with quantitative data or facts. We thus conducted two questionnaire surveys aimed at providing data that would generate rigorous findings. Questionnaires were sent to all the companies and the executives that we had met, the results of which are found in Chapters 4 and 6 of this book.

Through our interactions with companies we observed fascinating examples of ESI success and failure, as well as product development efforts that exemplified what we had discovered about the requirements of successful implementation. We therefore wrote some in-depth case studies for use in management development seminars: Philips Floor Care's Triathlon project, Lexmark's 'Liberty' printer project and the Fuji Xerox

DocuColor photocopier. The last of these case studies was completed in 1997, following a research project conducted at the Theseus Institute. In this book, we provide summarized and partial versions, often focusing on only one of the core issues in these cases.

The surveys and case studies alerted us to the long-term implications of ESI. Involving suppliers in product development is not without consequences in how companies allocate responsibilities among the industrial value chain. Companies facing ESI decisions are increasingly anxious to understand the longer-term implications. We are thus now exploring this issue and even though our research in the area is still developing, we will share some early findings in the later chapters.

The book starts with an explanation of why Early Supplier Involvement has become a 'burning' issue, and does so by reviewing the major questions in today's product development and sourcing. This concludes with the definition of a leverage mechanism that uses the competence and creativity of suppliers to extend the innovative level of their clients.

The definition of Early Supplier Involvement is not entirely straightforward. As we discuss in Chapter 2, there are different degrees of ESI. We contrast ESI with more traditional ways of interacting with suppliers for product development through a series of cases. We also report the benefits that clients, as well as suppliers, have found in this new approach.

The story of ESI as a management innovation offers some insights in understanding its real nature. There is little doubt that ESI originated in the Japanese car industry, and especially at Toyota. The reasons behind this, however, are certainly not as well known ... nor is the process that diffused ESI across continents and industries. In the third chapter, we discuss the roots of ESI and outline its diffusion and evolution.

Chapter 4 reports on the findings we drew from our questionnaire survey of ESI clients. We discuss the characteristics of ESI adopters and identify those that are most associated with the practice. One message comes out loud and clear: ESI is primarily a matter of strategic choice. It is clearly implemented by companies that have previously put a number of policies in place regarding issues such as supply base management.

But ESI is not a smooth journey for executives who set out to implement it. They will typically be faced with reluctance on the part of some colleagues. There are, indeed, as many reasons to reject ESI as there are to adopt it. In the fifth chapter, we attempt to discuss the most common (16!) objections to ESI that we encountered and we offer responses based on our research.

Chapter 6 turns to the suppliers' perspective. It shows that if clients enjoy benefits through ESI, these are not obtained at the expense of their suppliers:

many, in fact, have a positive attitude towards ESI. Some, like Nypro, Minco and Seitz, actually promote ESI as a way of differentiating themselves from their competitors. But this clearly requires new types of relationship management and supply contracts.

Chapter 7 discusses a key concept in many managerial situations, and in ESI in particular: trust. Trust is essential in managing the interdependence that characterizes ESI relationships between buyers and suppliers. But although it is now *de rigueur* as a remedy for many business problems, it is poorly understood and even ignored by most economists. Borrowing from the most recent discussions in this area, we submit a new definition of trust and its source which we believe will help executives manage ESI relationships more effectively.

The last chapter offers a synthesis of our ideas. It outlines five key implications for the strategic management of ESI: (1) manufacturing companies will increasingly have to cope with the end of their own self-sufficiency, (2) the new strategic drivers will emphasize the ability to deal with time pressure and complexity simultaneously, (3) effective knowledge management will become a centrepiece in this context, (4) the organization will continue to move away from traditional forms and towards connected networks, and (5) innovation will rise up to the top of the strategic agenda.

The chapters in this book do not need to be read sequentially. The busy executive who is preoccupied by implementation issues will find Chapter 5 pertinent. We suggest to these readers that Chapters 2 and 3 will add to their understanding of the issues. Chapters 1, 4 and 6, even though they may seem academic, provide a footing for the policy decisions that must be set for ESI to succeed. Finally, Chapter 8 offers a comprehensive view of the strategic context which will be meaningful for those who understand the nature of ESI and want to explore its implications.

This book is the result of a substantial effort and would not have been completed without the support of many individuals to whom we would like to express our gratitude. First among them is Professor Thomas Vollmann who, as we previously mentioned, offered his support for the initial research that now results in this book. Similarly Professor Robert Collins, who succeeded Tom as Director of IMD's Manufacturing 2000, deserves heartfelt thanks for his outstanding help through the early stages of our research. Our colleagues at the Theseus Institute, which two of us joined in 1996, deserve our gratitude for their friendly support.

Since the heart of this book is based on an extensive survey, we are indebted to the many executives who accorded us some time in spite of their busy schedules. It is impossible to mention them all here. However, we would like to thank in particular the people and organizations who helped us access high-level executives in the US and in Japan. Patricia Moody of the AME (Association for Manufacturing Excellence) was very effective in introducing us to prestigious American companies from which we learned a great deal. In Japan, the JRIA (Japan Research Institute, Limited) and our JRIA contact Mr Eiji Hayashi, President of NEC Patent Service, were extremely supportive and friendly. Without their help we would not have had access to many fascinating companies. We also want to express special gratitude to the executives who helped us write the in-depth case studies that added so much to our understanding and, hopefully, to that of the reader. They include Wil Van de Berg, Carel Taminiau, Hugo Van de Woord, Henk de Jong from Philips Floor Care; Greg Survant and Joe Walden from Lexmark and Gary Deaton from Lexmark's supplier Minco; Mr Toshiyuki Iijima and Mr Tatsuo Kobayashi from Fuji Xerox, as well as our colleague Professor Hirokazu Kono from Keio Business School. Most of them have become friends in the process and we hope there will be other opportunities to work with them in the future.

The traditional disclaimer applies here. In spite of the support given to our successive research projects, we obviously carry full responsibility for the mistakes, approximations and limitations contained in this book. We nevertheless hope that the reader will enjoy a text which took much time and triggered much passion as we produced it.

1 The Context of Early Supplier Involvement

'A car for two people and a pack of beer.' This is how critics describe 'Smart', the micro-car due to be launched in the very near future and expected to sell at a rate of 200 000 by the year 2000. The Smart car departs from conventional automotive design logic – it is nothing, in fact, if not innovative. This new product, aimed at frayed nerves and urban congestion, is 40 per cent shorter than most European sub-compacts and capable of being parked perpendicular to sidewalks. A top-of-the-line 'hybrid' model is equipped with both a gas engine and an electric motor. The body of the car is fitted with plastic panels that can be changed in a short time to give the car a new look in a flash. Smart's marketing is also expected to break new ground with a novel approach to leasing: consumers will have the option of leasing a 'transportation solution' where, for a monthly fee, they buy not only the Smart car but also the right to use a large sedan for a period of time each year.

Critics note that the projected price (US $10 000 or ECU 8500) puts Smart in competition with Fiat, Renault, VW and others who offer sub-compacts that, for a similar or even lower price, permit long-distance travel on motorways, can carry heavier loads, and also navigate city traffic. Some financial analysts question sales forecasts, which they feel are too 'optimistic', and believe investors are taking risks. One analyst considered that sales might be as low as half the official Smart projections.[1] Whether or not Smart will ultimately succeed is an interesting issue. More interesting, however, is the way the Smart project was organized. This new product development effort illustrates particularly well the profound changes that are redefining corporate boundaries.

Smart is the product of MCC (Micro-Compact Car), a joint venture formed between SMH (the Swiss company that invented the Swatch watch) and automaker Daimler-Benz. The initiative came from SMH under the leadership of CEO Nicolas Hayek, the man who orchestrated the recovery of the Swiss watch industry after it had lost ground to Japanese watch makers. Dr Hayek invented the Smart concept but decided early on that SMH could not go it alone for lack of automotive experience and know-how. He therefore approached several mainstream manufacturers and proposed a joint-venture arrangement. Volkswagen was one of the first companies that SMH contacted, but Hayek could not convince VW to enter into a cooperation agreement. He then turned to Daimler-Benz, the German manufacturer of up-scale cars that has suffered from Japanese inroads into the luxury car market. Daimler-Benz's new strategy no longer limits the company to large luxury sedans and fleet cars; instead, its product lines are extending into new segments that include compacts and minivans. The Smart concept was a good fit with this strategy, and Dr Helmut Werner, the Chief Executive of Daimler-Benz at the time, decided to enter the joint-venture arrangement by taking a 51 per cent stake in MCC to SMH's 49 per cent. The ownership distribution between the shareholders has changed several times since then.

Joint-venture arrangements are common in the car industry. Manufacturers often rely on JV structures to share investment costs related to the production of components (PRV, for example, which was formed to supply engines for Peugeot, Renault and Volvo) or even entire cars (SEVELNORD, for example, which manufactures minivans sold under four different brands owned by Fiat and Peugeot). Generally, these JVs are built around relatively mature products and technologies, permitting partners to benefit from scale economies and cost sharing. The Smart project adds the dimension of innovation – some would say radical newness. An Andersen Consulting partner, closely involved with the project, was quoted as saying, 'Smart is a revolutionary car with innovative ideas, teams, and processes producing it and bringing it into an exciting market.' In this venture, Daimler-Benz and SMH joined forces to innovate by combining SMH creativity with Daimler-Benz technology and reputation.

Another striking feature of the Smart project is its heavy reliance on numerous companies to realize product development, production and distribution. Suppliers like VDO (cockpits), Krupp Hoesch (rear axles), Bosch (electronics) and Dynamit Nobel (plastic panels) became involved early in development discussions and were asked to invest in a greenfield manufacturing site in the town of Hambach, France. Some 60 per cent of this investment has been committed by the supplier base, representing a new venture in its own right. Suppliers are increasingly encouraged to invest in the vicinity of existing car assembly plants but the Smart project went a step further by creating an entirely new site. Supplier commitment includes a rumoured US $300 million in new manufacturing units and over half the employment at the Hambach site. In essence, these suppliers share the risks with MCC in the hope of sharing the ultimate rewards. And according to analysts, the risks were higher than a typical new automobile plant which generally supplies a well-known market segment with an existing concept.

The concept of partnership was also applied to distribution with the establishment of a dealer network specifically dedicated to Smart. Whether or not they were previously associated with Daimler-Benz, dealers will have to set up all-new dealerships – of which 100 were planned in urban centres throughout Continental Europe. The cooperative approach between MCC and its suppliers and distributors was even extended to consultants. In February 1997 Andersen Consulting was handed major responsibility for programme management and IT operations. But in the partnership spirit indigenous to MCC, Andersen will not receive its usual consulting fees; instead, it will be remunerated in proportion to Smart's success.

The Smart project exemplifies two profound changes affecting industry today: a drive to innovate through new products and processes, and the need to combine the resources of several companies in order to do so. Smart illustrates the changing nature of innovation processes which are increasingly managed through a strategy of partnership: independent organizations contributing different resources, with uneven degrees of involvement, yet each sharing the risks of the venture. This project illustrates the extent to which a strong industrial vision can leverage the resources and competencies of other organizations. Moreover, MCC is an example of how one manufacturer combined the competencies of other firms, suppliers, distributors and manufacturers to launch a radical product innovation.

Independent of its success or failure, the Smart project raises a substantial number of issues. This chapter will outline the main themes which subsequent chapters will address in greater detail. The following two sections discuss the origin and nature of two major industry trends illustrated by the Smart project: product innovation and supply base management. These two trends are leading to a new model of technology management where a key aptitude will be the leveraging of resources owned by other organizations, and especially suppliers, in order to innovate. We term this phenomenon **leveraged innovation**. This chapter reviews the managerial implications and raises a number of issues that are confronting executives today.

Product Innovation: The Changing Agenda

Innovation has long been celebrated as a vital activity of mankind. Economists herald it as an essential source of economic growth and wealth generation; engineers underscore its value in providing firms with new products and processes; entrepreneurs look upon it as a source of profits. The intended, formalized gains are numerous, especially since the advent of the capitalist era, and particularly today when the benefits of technological innovation go largely unchallenged. Both academics and managers now rely on product innovation and expect rewards that exceed those previously experienced. But as we will explain, there is growing awareness that innovation does not occur in a vacuum. Research provides clear evidence that innovation requires, and will increasingly require, cooperation among firms. These two ideas form an interesting paradox: innovation is critical to a firm's success but increasingly requires interdependence with others.

The high up-front costs of today's technology development, as well as the variety of competencies required, make it nearly impossible for individuals to undertake innovative activities on their own. Thus, most innovations now develop within organizations and product innovation has become a major competitive weapon in some industries. The strategic dimension of product innovation can be understood on two levels: as an output of the creative process that enables a firm to compete effectively, or as a means to actually reshape the rules of the game. We will review these two dimensions and discuss a newer approach to innovation that presents it as a change process, embracing both organizational learning and competencies.

Innovation and Competitiveness

The first level will be termed 'product creation' following Nayak and Deschamps[2] and occurs where competitive advantage is obtained through better value and/or lower cost due to superior design. In this respect, product design is an essential step in gaining or retaining competitive advantage because the key dimensions of competition, value and cost, are determined at the design stage. This is obvious concerning a product's value dimension since product performance is established primarily at the development stage through the linkage with specifications. For example, in the computer printer market the speed (number of pages per minute) as well as print quality (dots per inch) are characteristics that make a product more or less competitive given a certain price range. Desktop printer manufacturers such as Hewlett-Packard and Lexmark are engaged in an ongoing rivalry to offer faster and sharper printing capabilities, while the market price follows a steep downward curve. To this end they have both introduced new technical solutions, such as ink jet printing, and continuously perfected existing products.

The cost dimension is also largely determined at the development stage. In the watch industry, for instance, it is estimated that 80 per cent of a model's cost is frozen at the design stage, implying that even the most efficient manufacturing systems can affect only the remaining 20 per cent. Similarly, development decisions in the desktop printer industry set the cost boundaries of a new product. Lexmark understood this well when they undertook the development of their low-cost 4037 printer in 1991. They put the target price below US $1000 in the product brief and this significant challenge was achieved, as we will explain in Chapter 4.

Innovation as a Way to 'Redefine' the Industry

Product innovation does not simply offer superior or cheaper products; it may actually redefine the competitive game. James Utterback, who pioneered research on the dynamics of industrial innovation with William Abernathy in the 1970s, has discussed the impact of innovation on industry competencies.[3] He explains that new products and new manufacturing processes can either build upon existing competencies or destroy them through obsolescence, by introducing an innovation that requires different skills and competencies altogether. There is evidence that challenging conventional wisdoms can be effective. A recent *Harvard Business Review* article[4] reported a research programme that compared firms following conventional product thinking with firms that did not. The latter did not consider industry conditions as given but rather believed they could be shaped. Nor were they content to match competitors' performance, but instead aimed to achieve 'a quantum leap in value to dominate the market'. These challengers pursued what the authors termed 'value innovation' and reaped the benefits: they represented only 14 per cent of the product launches in the sample but generated 38 per cent of the total revenues and 61 per cent of the total profits.

The authors, W. Chan Kim and Renée Mauborgne, illustrate the 'value innovation logic' through the example of Formule 1, a French hotel chain pioneered by Accor in the 1980s which took a radically different approach to the product/service bundle traditionally offered by one-star and two-star hotels. Formule 1 hotels have virtually no receptionists, small rooms, no eating facilities or lounge space, but match two-star hotels in room comfort and hygiene and one-star hotels on price. 'Formule 1 made the competition irrelevant' and this gave them a market share greater than their five largest competitors combined. Other value innovation benefits implicate the losses experienced by conventional players. Utterback has used the example of typewriters to show that major innovations, such as the advent of the electric typewriter and PC-based word processing, have undermined market leaders like Remington who were unable to evolve their competencies fast enough to match the competition.

Innovation Activities as Change Process

If innovation is critical in producing competitive products, academic research shows that it also plays an important role in ensuring that companies maintain, or even improve, internal business processes. New product development is more than just product creation: it can be an essential feature of a company's change and development process. When product innovation alters the conditions of competition, as Utterback explains, it does so by affecting the capabilities of incumbents. Dorothy Leonard-Barton[5] has addressed the organizational implications of this mechanism. She developed a rigorous definition of capabilities (including technical systems, managerial systems, skills and knowledge base, and values and norms) and then underlined the value of 'core' capabilities for product innovation. While existing manufacturing expertise might plot a development vector for future products, a firm's core capabilities could – in the face of competitive developments – prove to be 'core rigidities' if the organization insists on applying old solutions to new problems. As the saying goes, 'if the only tool you have is a hammer, everything looks like a nail'. Project leaders thus face an ambiguity: they need to take advantage of existing capabilities without being constrained by them.

The goal is to ensure that core capabilities evolve should the competitive environment so require. Leonard-Barton explains that this is precisely where development activities play a central role. New product and process developments are unique opportunities to nurture new organizational capabilities that, although they may call traditional 'core' competencies into question, could offer alternatives as market conditions change.

But for new capabilities to become core, they must be integrated 'deeply' within an organization. Simply adjusting existing technical or managerial systems is insufficient. Skills and values must evolve because 'a core capability is an interconnected set of knowledge collections – a tightly coupled system'. In her 1995 book, Leonard-Barton[6] explains how this can be achieved. She identifies several activities that are critical to developing a firm's capabilities: shared problem solving across cognitive and functional barriers, implementation of new methodologies and process tools, experimentation and importing know-how from sources outside the organization in order to extend knowledge.

Innovation and learning are intimately related; Ikujiro Nonaka and Hirotaka Takeuchi have argued that they are two sides of the same reality. In their book *The Knowledge-Creating Company*[7], they describe how leading companies in Japan (for example, Honda, Canon, Matsushita) manage innovation through the formulation and circulation of knowledge. In particular, learning is created through the interaction between tacit and explicit knowledge in organizations. This interaction, properly guided, results in an accumulation of knowledge from which innovation opportunities arise. For instance, the 'on-site' observation of a baker's craft by a Matsushita engineer and her subsequent description and analysis resulted in new knowledge that led to the successful development of a bread-making appliance. Knowledge is thus the soil in which innovation grows.

But, while new products follow their life cycle, knowledge accumulates continually. This process, according to Nonaka and Takeuchi, is not limited to the technology and engineering functions since innovation typically involves large numbers of staff and departments. In particular, the role of middle management as the link between top management's vision and operational reality is underlined. Middle managers are essential to the learning process and the innovative potential of an organization.

At a time when the key assets of a firm are not limited to financial wealth, but include the scope and quality of the knowledge base, product innovation appears to be an expression of organizational expertise as well as a source of knowledge. Firms can support organizational learning through innovation activities and the resulting accumulation of knowledge encourages the creation of new products and processes. This, in turn, fuels an interesting dynamic that combines innovation, change and growth. The 'social' dimension of this dynamic is quite clear as it necessarily involves actors outside the organization.

Innovation as a Social Activity

Innovation is often portrayed as an act of freedom *par excellence*, the achievement of an ingenious individual who challenges the status quo. We are all familiar with great innovators like Honda or Pasteur and would like to believe that they worked alone, triumphing against the odds. This image of the lonely innovator is, however, firmly rooted in the Western apotheosis of the 'individual' and not supported by hard evidence. There are certainly examples of major innovations accomplished by individuals but more often, new products have been developed through the coordinated efforts of several people. Product innovation is largely a social, collective process. Even when the kernel of an innovation is introduced by an individual, development and commercialization typically involve many others. This is clearly demonstrated in research concerning the determinants of innovation effectiveness: the role of 'outsiders' is present in the four main research streams constituting this field.

The 'Key Success Factors' of New Product Development

A 1992 article in *R&D Management Journal*[8] reviewed several decades of research on the conditions for success in new product development. The author, Ian Barclay, found no fewer than 23 different papers published between 1956 and 1990 addressing this question. Generally, the first line of research consisted in looking for the underlying characteristics of success in a sample of development projects. Overall, 140 different factors were associated with success in these various research efforts. Despite this number, the findings are remarkably consistent. Table 1.1 presents a

Table 1.1 The most common key success factors in product development (*Source:* authors' synthesis of literature in this domain)

1. Product uniqueness, relative to customers' needs
2. Marketing and distribution competence of the developer
3. Excellent up-front planning in early phases of development
4. Ability to develop or acquire appropriate technology
5. Presence of a product champion and sponsorship from top management
6. Use of appropriate control, different from the type applied to 'mature' activities
7. Easy and fast communication in the firm

synthesis of the most common characteristics of successful product innovation programmes.

While some of the key success factors involve the product itself, others implicate project management and still others the organization that houses that project. The factors related to project management make it clear that the presence of a champion and a sponsor is important, along with the need for appropriate control mechanisms. The importance of a firm's ability to develop or acquire appropriate technology is documented by a large body of research on product development effectiveness. Failures[9] have been associated with the insufficient status of technologists within the firm (Carter and Williams, 1956), a lack of R&D resources (Myres and Marquis, 1969), inefficient development work (Rothwell, 1977), poor synergy and proficiency of technology and production (Cooper, 1979), and a poor fit between the technology in question and a company's overall level of expertise (Souder, 1987).

This underscores the fact that successful product developers are able to match product requirements with the technology required. But some analysts (for example, Rothwell, 1977) suggest that innovators have to understand when technology should be developed internally, and when it should be acquired from another organization. At times, it is advantageous to nourish the product development process with outside sources. Just as Edison developed the electric light bulb with the help of Corning Glass, most product innovations benefit from the contributions of other firms. It is also clear that innovators often fail when they insulate the technology development process and exclude outside sources, regardless of their expertise. It thus becomes important to know what one is good at, and to find ways to address technology deficits in collaboration with outside sources when appropriate.

But even if some authors acknowledge the need for outside contributions, the literature says little about what firms should actually do to mobilize external development resources. This limitation is typical of the key success factor approach: as explained by Brown and Einsenhardt, 'fuzzy' concepts such as 'unique products' or 'excellent up-front planning' are of little practical help since they typically generate little more than a laundry list of concerns.[10] To make matters worse, Barclay found that very few new product managers were actually even aware of these results.[11]

Brown and Eisenhardt note that two other approaches shed light on the determinants of product development performance. The first focuses on the role of communication processes, and the second analyses the management of product development through teams and management units. Both emphasize the importance of external sources of information.

The Importance of Communication in New Product Development

Following the pioneering work of Tom Allen at MIT, a succession of scholars have tried to understand the impact of communication on development activities. Communication is a dimension of innovation management which, in the research stream mentioned above, has been identified as a key success factor. Generally, researchers observe that frequent and substantial communication improves development performance. More specifically, internal communication (among project team members) was found to be critical to task coordination across different departments and organizational functions. Research also showed that external communication (between the team and its environment) was useful for task coordination, but that the major benefit was securing necessary resources through 'political' and 'ambassadorial' activities which protected the team from external 'attacks'. This idea is consistent with the concept of a project sponsor who shields the project champion and his/her team from opposition, whether internal or external to the firm.

One particular role was identified by Allen in connection with external communication: that of the 'gatekeeper'. This is a team member who communicates with the outside world more often than his/her colleagues, then transmits the accumulated information to the team and thus improves its communication effectiveness. It was found that teams with a gatekeeper tended to perform better than those without one.[12]

Eric Von Hippel, also from MIT, demonstrated the importance of communication with customers, in particular the lead-users who are often a primary source of innovations.[13] He argued that this segment of the market generally experiences needs ahead of other segments because it is an intensive user of the product in question. In the powertool market, for example, some people are passionate woodworkers and more likely to look for new solutions than the average homeowner. Active communication with this type of user yields market insights and a broader understanding of potential innovations. This is how Bosch, for instance, developed its 'detail sander' which outperforms other power sanders in the market for specific tasks that require precision.

This line of research highlights the social dimension of innovation, both within and across organizations. It demonstrates that the innovation process is a boundary-spanning activity in which potential innovators need the input of others within the organization, as well as from other sources. This is an important contribution to our understanding of innovative activities. But the management of innovation across organizations requires more than effective communication, and this is where a third line of research, as defined by Brown and Eisenhardt, offers some interesting insights.

A Japanese Approach to Product Development?

This research stream originated with a series of clinical studies of Japanese new product development processes, which were conducted both by Japanese (for example, Imai, Nonaka, Takeuchi, Fujimoto) and North American academics (for example, Quinn, Clark, Eisenhardt). Singly and sometimes in teams, they focused on successful Japanese companies such as Fuji Xerox, Canon, Honda and Toyota in order to understand the 'Japanese model' of product development.

Interestingly enough, this literature associates the concept of success with speed or time to market. Brown and Eisenhardt also explain that the recipe for success includes ingredients that may actually conflict, creating interesting dynamics:

- An autonomous team responsible for problem solving
- The leadership of a 'heavyweight' project manager
- Strong top management exerting 'subtle control'
- A clearly articulated product vision

This model would appear to solve the traditional opposition between centralization and decentralization in product development. The development team is empowered but top management does not hand over total responsibility; instead, the mandate specifying development objectives is a short, clear message that delivers an essential concept upon which the team focuses.[14] Brown and Eisenhardt have termed this approach 'disciplined problem-solving': disciplined because of top management's mandate, and problem solving because decisions are made at the project team level. The approach also suggests a way to overcome the classic tension between project teams and functional departments through the concept of a 'heavyweight' project manager. This is a senior executive with sufficient experience, authority and political clout to achieve cross-functional integration. This model makes it possible to implement concurrent engineering to speed up the development process.

The Japanese approach featured another departure from traditional methods: the involvement of suppliers. Nowhere else than in the study of the car industry were the differences so pronounced. In their worldwide study of 29 projects among 20 car manufacturers, Clark and Fujimoto[15] showed that suppliers to Japanese car makers played a much larger role in product development than their American counterparts, taking charge of a larger portion of the overall development in terms of cost and engineering man-hours. The share of suppliers in total engineering effort for new car development was 30 per cent in Japan compared to 7 per cent in North America and 16 per cent in Europe.

But Brown and Eisenhardt caution that lessons from the car industry should not be generalized.[16] A recent study of the world computer industry by Eisenhardt and Tabrizi concluded that some of the project management features identified as favourable to effective product development did not seem to apply in the most dynamic market segments. For instance, '... computer-aided design, rewards for schedule attainment, supplier involvement, overlapping development stages, or extensive planning not only did not accelerate the pace, but, in fact, often slowed it'. The authors also concluded that the best approach to product development management depends on market conditions. In general, this line of research yields valuable lessons but also raises significant implementation issues. On the issue of supplier involvement, for example, most recognize the benefits of ESI but fail to discuss its implementation, even though this initiative requires an organizational change process away from a 'closed' innovation system to one where external contributions become central. Another line of research focuses on the social dimension of innovation through the communication perspective noted earlier, but adding the aspect of a firm's overall strategy.

A Network View of Innovation

Interpersonal relationships have long been identified as crucial to the diffusion of innovation (for example, Rogers[17]), and this is also the case in the organizational adoption of technological innovation. Engineers obtain access to technological information (such as the problems associated with the usage of a new technology) from friends or colleagues working with competitors or suppliers (for example, Czepiel[18]). We personally know of a French engineer working on the European Ariane rocket programme who overcame a technical difficulty with the booster by consulting a NASA engineer. In the absence of networked relationships it is probable that many innovations would never have been achieved. It would be a mistake to interpret such communication as a lack of loyalty; instead, it is an integral part of innovation strategy. Von Hippel, in fact, argues that engineers trade on their expertise – they provide what might be regarded as confidential information in the hope that the favour will be returned. This establishes a parallel community alongside the formal corporate structure, with its own behavioural rules and norms. The sociologists Callon and Latour made a series of major contributions in this area. Through in-depth clinical studies they showed how innovations need to be developed within a network of actors. This network is essential long before the diffusion stage: it can protect the development team, somewhat like the sponsor role noted earlier; it can provide resources and necessary knowledge; and it supports the diffusion of innovations. The fascinating story of the development of Aramis, a 'personal transit system' (Latour[19]), details the inner workings of the network of actors involved in this failed development. Latour showed that technology developments which, like Aramis, endeavour to preserve their 'technical purity' without factoring in the concerns of other actors in the network will never reach the status of true innovations – that is, socially accepted and marketable technological objects.

Industrial marketing specialists, such as Håkan Håkansson and his colleagues in the Industrial Marketing and Purchasing (IMP) Program project, also analyse innovation in terms of networks. Their approach is substantially different from others, especially with respect to methodology, and clearly supports the social concept of innovation. Given their focus on industrial/business-to-business goods, these authors insists on the 'interactive' nature of innovation. New products and new processes are said to emerge from continuous interaction between participants in a network (suppliers, clients, complementary products providers, distributors): the cooperation between these actors allows them to 'mix' knowledge (for example, material science and mechanical engineering) and to take into account their economic interdependence. In this context, technology development cannot be understood as a rational process but rather as an exercise in political judgement, diplomatic skills and social interaction.[20] The managerial implication of this line of research is that openness to outside contributions is an important condition for successful innovation. This is valuable in terms of the implementation issues raised by the social nature of innovation and we will return to several of these concepts later.

Despite differences in emphasis, the four research streams outlined here all conclude that product innovation is a social process. Moreover, the most recent research (the 'disciplined problem-solving' and 'network' approaches) insists on the importance of external contributions. We believe that with the passage of time, the social dimension will become a prominent factor in the effectiveness of innovation processes. In the next section, we discuss some apparent drivers behind this trend.

The Drivers Behind the Socialization of Product Innovation

There are many factors driving this evolution but it is not our intention to be exhaustive. We will revisit the issue when we examine the drivers behind the involvement of suppliers in product development as some of these drivers are transversal. Instead, we will focus on two factors that play a fundamental role in making product innovation an increasingly social process. The first, product proliferation, results from empirical data while the second, technology fusion, is rooted in the theory of innovation.

Product Proliferation

This first trend is evident in most industries and especially among assembled product markets. It consists of a continuously growing number of product categories offered by incumbents, either as a result of their own initiative or as a response to the offering of a new entrant. The snow-ski industry offers a good illustration. Dating from its mass-market beginnings during the 1950s, ski manufacturers have offered a range of alpine skis. A few innovations were introduced through the 1960s and 1970s (new materials such as steel and plastic, and torsion box structures), but the design of the downhill ski remained basically unchanged. During the late 1970s and early 1980s, manufacturers incorporated Nordic skis in their product lines and the largest firms diversified into ski boots. But the 1980s and especially the 1990s have witnessed new concept introductions: the monoski, the 'big foot', the snowboard, the carving ski, as well as new production processes such as pre-impregnated composite materials.[21] The net result is an expanding product line corresponding to a fragmentation of winter sports activities.

In an industry like snow sports, it can be difficult to distinguish durable market changes from fashion swings. But when examining an industry less subject to fashion, such as vacuum cleaners, we find a similar trend. Interestingly enough, the concept of a vacuum cleaner varies substantially across countries but in the European market, the two dominant types of vacuum cleaner are the 'upright' (which carries the central unit on a vertical arm) and the 'sledge' or 'cylinder' (with the central unit rolling on the floor and connected to the nozzle via a flexible pipe). New vacuum cleaning products have been introduced over the years, including cordless machines, wet and dry machines and more recently, the injection and steam cleaners.[22] Generally, these new products are simply added to an evolving line of existing products without an obvious substitution process. Consequently, manufacturers find themselves managing an ever-growing product line, which puts a strain on engineering capabilities.

Continuous product line extension has been used as a strategic weapon to exhaust competitors, a move defined as 'product proliferation'. Sony, for example, is expert at deploying this strategy and a visit to Sony's Walkman museum in downtown Tokyo provides ample evidence. The number of different products is impressive in and of itself, but the variety of designs introduced every year is literally amazing. Sony attacks smaller segments of the market each year, developing the Walkman for every consumer need and listening situation. This trend was recently reinforced by the advent of digital recording technology which increases the number of possible product permutations. Technological innovation is thus combined with market segmentation to multiply product opportunities.

Technology Fusion

It is commonly accepted that innovation results from the cross-fertilization of different fields of knowledge. This is perhaps what led Professor Fumio Kodama, a Japanese Policy Science Institute academic, to document the concept of technology fusion.[23] Studying technology management in leading Japanese manufacturers, Kodama observed that in contrast to their American counterparts who generally focused on an industry's dominant technology, Japanese high-technology manufacturers fused different technologies together in an attempt to create significant innovations. Kodama cites NTT, NEC and Sumitomo Electric Industries which combined 'glass, cable and electronics technologies to produce Japan's first fiber optics' and who, today, are the founders of 'optoelectronics' in which they enjoy commanding market shares. Kodama also notes the case of Sharp which fused crystal, electronics and optics technologies, and is now dominant worldwide in Liquid Crystal Displays.

Kodama argues that the pursuit of breakthroughs by focusing specialization on one area of science or technology does not have the same innovation potential as that offered by fusion. He speculates that Western firms may be reluctant to exploit the benefits of fusion due to 'a distrust of outside innovations, a not-invented-here engineering arrogance, an aversion to sharing research results' (Kodama[24]). The fundamental blockage appears to be a discomfort in working across boundaries, and yet fusion requires organizational openness to gather and exploit technology developments that might have potential applications. Openness is also implied through the ability to manage effective R&D relationships with various organizations that possess complementary knowledge. Contrary to what one might think, this attitude is not technology-driven; rather, it stems from a profound understanding of market demands. Kodama explains that the key to this process is 'demand articulation', that is, the translation of often ill-defined consumer needs into long-term product and technology requirements.

According to Kodama, '[technology] fusion will play an increasingly important role in product development efforts in the future'.[25] He foresees the growing interpenetration of technologies across industrial sectors which will lead firms to interact (in terms of their R&D activities) with suppliers from previously unrelated industries. An illustration is the development by Salomon of the Monocoque ski, whose design included the use of pre-impregnated plastics, a special composite material with glass fibre and resin that remains in a state of partial polymerization, and originally developed for aeronautics applications (for example, helicopter rotors at Aerospatiale). This development, which required Salomon's cooperation with outside material specialists, led to one of the most remarkable innovations in winter sports and contributed to a restructuring of the ski industry during the 1990s.

Product proliferation and technology fusion reinforce each other. The former sometimes requires the mobilization of new technology, such as the new vacuum cleaners that incorporate new functions which require new technical solutions. The latter, on the other hand, creates opportunities. Digital broadcasting, with its microprocessor and software technology, rejuvenated the television equipment industry with the production of 'set top' boxes that supplemented existing product lines.

The social dimension of product innovation generates a set of issues for manufacturers. It requires the management of product and technology development by coordinating with other organizations in different industries, the government or education. This departs from the view of the innovator's solitary genius but raises issues that innovators will confront in the future. One is the nature of innovation relationships: could the innovator not simply buy the technology that (s)he needs, just as Edison bought the piece of common cotton that he carbonized to make the first light-bulb filament? In this case, the market could actually deliver the needed material because it already existed. But what if the needed material or technology does not exist? Innovators increasingly confront this situation: their innovations are based on another's innovation, a rather typical situation of uncertainty that market relationships address with difficulty. For this reason, innovation processes increasingly require that organizations abstain from traditional commercial relationships and engage in medium- to long-term collaborative arrangements.

This presents an intriguing paradox: on one hand, product innovation is an essential element of competitive advantage while on the other, it cannot be achieved independently. In essence, competitiveness is achieved by cooperating with other firms. If this is the case, how are the fruits of competitiveness – profits in particular – shared with other firms and in what ways? How is ownership of the innovation attributed? When regular supplies follow the product's development, what logic governs the transactions? How will transfer prices be established? It is clear that a number of important issues are raised by this approach to product innovation. Related trends are now observed in manufacturing industries, most notably changes in the way firms are managing their suppliers.

Supply Base Management

An Increasingly Interdependent Connection

Changes in supplier–manufacturer relationships during the late 1980s and early 1990s have occurred more abruptly than the evolution in product innovation processes. While the latter evolved since the advent of capitalism and progressively reinforced the social dimension, purchasing processes appear to have undergone a major change within the span of a decade. Following the lead of firms that are deeply engaged in global competition, a growing number of companies have launched drastic reviews of their supply base management policies. This restructuring, which we will treat in the next section, has led to considerable discussion on the future of buyer–supplier relationships. The academic discussion of this phenomenon (which followed more than it led practice) focused on manufacturing initiatives to reorganize the supply base. In the following section we argue that despite a variety of interpretations, the main implication is increased strategic and organizational interdependency between manufacturers and suppliers.

From a purely descriptive standpoint it is clear that executives are talking and acting as if they now have a new relationship with their suppliers. Phrases like 'partnerships', 'close links' and 'joint destiny' are commonly employed. But before qualifying these terms and relationships, it is useful to review the empirical data regarding actual changes. In so doing we will see that the changes form an interesting pattern: an increase in outsourcing, based on fewer suppliers who, in turn, are carefully selected and accredited, and more tightly linked to their client(s), who agree to support their efforts.

A Broader Scope of Outsourcing ...

Business Week has hailed outsourcing as the 'growth industry of the Nineties'.[26] It is clear that a growing number of companies are engaging in outsourcing: 86 per cent in 1995 versus 58 per cent three years earlier. More importantly, the scope of outsourcing is broadening. In the 1990s, outsourcing was practised in areas traditionally considered an internal responsibility, such as customer-service calls, training and development or network management. Some companies, such as Gillette, began to question long-held convictions about manufacturing activity; while plastic moulding had always been an in-house activity at Gillette, the razor giant decided to entrust Nypro, a leading US plastics processor, with their successful women's razor. No activity seems unaffected by this trend. Bose, a leading US manufacturer of high-performance audio speakers, has outsourced some of its purchasing function to specialized suppliers who participate in its 'JIT-II' programme.

... Based on Fewer Suppliers

At the same time, manufacturers claim they want to reduce their supply base, and this may not seem very consistent with the first trend. Examples of supplier reductions are plentiful: Xerox stated it had ten times fewer suppliers in 1994 (400) than in 1984 (4000); Diebold, the US leader in ATM equipment, trimmed 850 companies from its supply base between 1989 and 1994 (1250 to 400) and intends to pursue this trend; Whirlpool cut its supply base of 2500 in half through a five-year crash programme; European Philips Car Systems (car audio) has implemented an ambitious plan to reduce suppliers from 350 to only 50. The combination of these first two trends is not inconsistent. It implies that traditionally outsourced activities will occur through a diminished number of suppliers.

... Carefully Selected and Accredited

The remaining suppliers will therefore do more, including subassembly and design. This is why most manufacturers now have purchasing programmes that limit orders to the most capable suppliers, and these are identified through an accreditation process. According to some observers, the ISO 9000 fad was fuelled by the spread of accreditation expectations among industrial clients who insisted that their suppliers be certified. But few companies understood the implications of this demand: failing proper definition of expectations and the appropriate support, supplier selection has limited value. Motorola's QSR programme or Honda's B/P programmes are good examples of initiatives that went beyond the fad and dealt seriously with the initiative.

... And Tightly Connected to Clients

The trend toward fewer but advanced suppliers is paralleled by an evolution toward integration between clients and suppliers – a closer coordination of their principal activities. This is evidenced by a series of purchasing policies. First, the concept of single sourcing is increasingly practised. Many manufacturers now buy a given part from a given supplier, instead of through several competitive sources. Daimler-Benz, for example, turned to single sourcing in the early 1990s in order to reduce purchasing costs by concentrating orders on the most efficient suppliers. The company claims this action delivered DM 2 billion in cost reduction. 'Sole sourcing does seem to be common practice', wrote Richardson in an article on the Japanese automobile industry.[27] But this does not mean, as he explained, that a given supplier will monopolize a part; rather, clients privilege a few suppliers for each component but select only one for each car model. In this approach, which Richardson calls 'parallel sourcing', the selected supplier is expected to be the sole source over the duration of the model's life cycle. This is expected to provide clients with benefits in terms of quality and JIT delivery because coordination with the supplier can be better integrated in the supply chain.

The second policy change is a move from the subcontracting of parts to subassembly. As suppliers become more competent, clients hand over larger responsibilities. Nypro, whose core competence is injection moulding, undertook new activities such as joining, bonding, coding and decoration on behalf of its client Motorola. In addition to manufacturing parts, Nypro increasingly performs subassembly work with or without the help of other companies. This trend is inspired by the structure of Japanese automotive subcontracting with its now famous tier system: a hierarchy of suppliers where the most advanced interact directly with car makers, and itself subcontracts to a chain of smaller firms.

The closer coordination required by a supplier's increased responsibility has sometimes called for geographic proximity of production sites. This was also initiated in the car industry, with clients requesting 'co-location'. Other industries have followed suit to make this another trend. At GE's Louisville, Kentucky facility, for example, most suppliers are within a few miles of the industrial site. Nypro, as a lead plastic moulder, has confronted this demand so many times that the company has turned it into a competitive response, claiming to be the McDonald's of custom plastic moulding with its consistent quality and proximity to clients. Research has shown that proximity does improve logistics (such as inventory management through timely delivery) and therefore cost-effectiveness.[28]

Face-to-face communication is the last and perhaps most important coordination mechanism that we will mention here. Again, research by Jeffrey Dyer shows that direct communication enhances quality: as distance between clients and suppliers in the car industry decreases, so does the number of defects.[29] Toyota, which had almost 7250 annual man-days of face-to-face contacts with its suppliers, experienced on average 25 per cent fewer defects per 100 vehicles produced than Nissan, which totalled 3350 man-days of contacts and 40 per cent fewer defects than American car manufacturers, who totalled around 1000 man-days of direct communication. In addition, face-to-face contacts are shown to be inversely correlated to development lead-time. It is not surprising that manufacturers insist on proximity with their suppliers and demand the secondment of guest engineers to participate on a full-time basis in their development activities.

... Who Claim They Provide Support

Finally, there is a trend toward supportive attitudes on the part of clients in order to enhance the capabilities of their suppliers. Because manufacturers expect a change in a supplier's role (for example, moving from parts manufacturing to subassembly and possibly coordination of sub-subcontractors), some realize that suppliers need to develop their competencies and that this requires the client's involvement to ensure value chain convergence. Several large companies, again led by the car industry, have implemented comprehensive plans to improve their supply base, such as Honda, Motorola and Xerox. Philips Car Systems has devised an interesting programme of supplier development to improve performance, particularly in terms of productivity or quality. It begins with a one-day diagnostic aimed at identifying improvement opportunities and setting performance targets. The supplier agrees to share programme benefits with Philips, then follows a series of two-day workshops where Philips specialists work with the supplier's staff, and possibly the second-tier subcontractors, to implement new processes and methods. The benefits sharing is subject to contentious comments, as some suppliers view these types of programmes as a trick to obtain price concessions. Other suppliers such as Seitz, a medium-sized US plastic moulder, acknowledged that a programme like this, implemented by Xerox, was instrumental in getting them 'out of trouble'.[30]

Even if this new type of buyer–supplier relationship appears common across countries and industries, it would be a mistake to consider it a universal solution. Several authors have issued cautionary statements, stressing that tight coordination with suppliers is expensive in terms of the managerial time it consumes. In fact, Toyota and Nissan, who are often mentioned as models of supplier partnership, do not treat all suppliers the same way.[31] They consciously select suppliers with whom they seek partnership and manage the rest through 'arm's-length' contracts. The latter applies especially to commodity/standardized, open architecture and stand-alone parts or components. But for the most advanced supplies (customized, non-standard, closed architecture, integrative) the trends hold, though at different degrees across industries and regions. This type of relationship has been observed on the Western industrial landscape since the late 1980s and early 1990s, and discussed at length in the business and academic communities. Questions have been raised on the origin, potential benefits, differential benefits (buyer vs. supplier), and long-term economic implication. But the most fundamental discussion revolves around the meaning of this new type of industrial relationship, and in the following section we will review the issues involved.

A New 'Model' of Buyer–Supplier Relationships?

It is clear that the evolution in client–supplier relationships marks a substantial shift from previous arrangements. The implications are not entirely clear, though it is possible to identify two perspectives on this phenomenon.

The dominant interpretation relies on terms like 'partnerships' and 'cooperation'. Richard Lamming, for example, discussed long-term changes in the worldwide automobile industry and identified a trend toward cooperative relationships.[32] In his view, the Western car industry evolved from a climate of adversarial relationships in which manufacturers pressured their suppliers to more cooperative methods, thanks to their Japanese mentors. The partial success already achieved results from the industrial context in the West which, unlike Japan's, is based on the assumption that suppliers are subordinate entities. Lamming foresees developments that go beyond the Japanese model, a stage he terms 'lean supply' which is characterized by cooperative relationships between equal partners. A similar point of view is expressed by Patricia Moody whose book bears the unambiguous title, *Breakthrough Partnering*.

Other analysts claim that supply chain integration does not necessarily require partnering. A report published jointly by the consulting firm A.T. Kearney and the Manchester School of Management (UMIST) takes a clear stand against the idea that partnership is necessary in these new buyer–supplier relationships. It notes that partnering has become a new orthodoxy advocating trust and open relationships between firms, and warns against assuming excessive risks in order to establish these new relationships in an 'altruistic' world where adversarial attitudes have supposedly disappeared. Instead, the report states that supply chain integration does not mean the end of power struggles between uneven parties.

If they differ in interpretation, these various approaches concur about the growing 'interdependence' between clients and suppliers. As shown in Figure 1.1, which ties together ideas developed throughout this chapter, manufacturers are increasingly dependent on their suppliers. This is due to market and industry dynamics that lead them to reconsider the structure of the value chain: on one hand, customers are extending broader responsibilities to their suppliers in order to focus on core competencies while on the other, they require 'customization' in order to maintain, or even enhance, product differentiation in the marketplace. Consequently, manufacturers need to have their suppliers invest in new assets while, at the same time, ensuring close coordination in order to avoid losing uniqueness in the eyes of their customers. This requires an investment in the relationship which reinforces manufacturers' dependency on suppliers' capabilities. Conversely, the requests from clients mean that suppliers must invest in new managerial methods and production processes to undertake subassembly work, in new facilities closer to delivery sites, and this obviously increases their dependency. Suppliers will therefore expect a

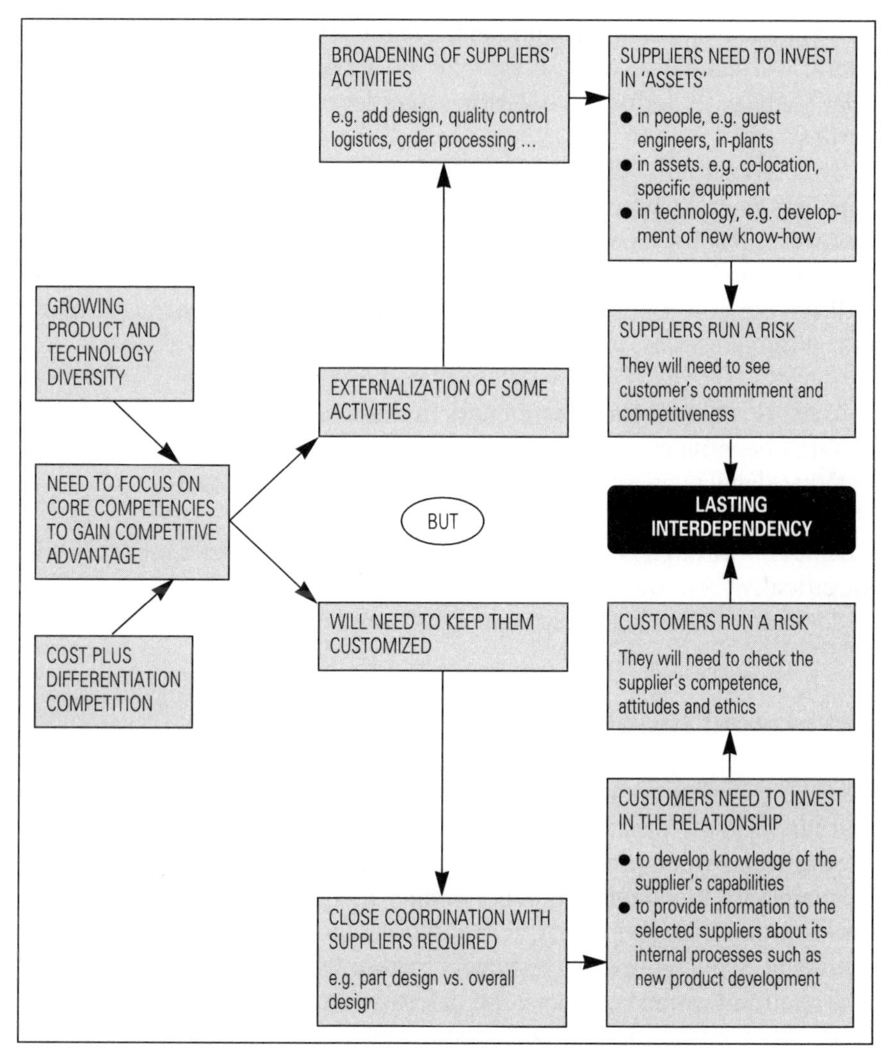

Fig. 1.1 The dynamics of buyer–supplier interdependency

client's commitment to the project at hand but because several clients may expect such investments, suppliers will tend to be selective and extend their support to the most attractive opportunities. Thus, a mutual dependency exists even though it may not be balanced. Furthermore, this dependency is long-lasting. To some extent a client has always been dependent on the supplier in terms of timely delivery or quality. But this dependency has been time-limited since a client could switch suppliers on relatively short notice. In this new context, the tasks assigned to suppliers are much more complex and the integration between two parties is much more extensive. Changing suppliers, even when it is allowed by contract, entails substantial efforts on the part of a client. We frame these new relationships in terms of a lasting interdependency between the two parties.

This constitutes the central challenge facing manufacturers and suppliers as we arrive at the new millennium. The issue is not dependence, because it has become a lasting attribute of industrial relationships. The central issue is learning to live with **interdependence**, managing it. We will devote considerable attention to this challenge which raises many intriguing questions, including:

- What is required for effectiveness in these new relationships?
- What benefits can be obtained from closer coordination?
- What dangers are hidden in the implementation process?

Chapters 4 through 6 will be devoted to such questions. Chapter 7 will specifically focus on the management of interdependence which requires, in addition to technical recipes, a significant level of trust.

Leveraged Innovation

Perhaps focusing on a specific aspect of the interaction between buyers and suppliers (on product innovation interactions) will clarify these issues. Product innovation provides an interesting perspective on buyer–supplier relationships. Conventional supply chain integration initiatives (such as co-location, dedicated plants, accreditation) can be interpreted in terms of dominant relationships where buyers require that suppliers assume more. The relationship can be understood in terms of a balance of costs: the client bears them, or the supplier bears them. What one wins, the other loses.

The meaning of collaboration in product innovation is substantially more complex. There is more ambiguity about who stands to win most – the buyer or the supplier. When both cooperate in the development of a new product, significant bipartisan gains are feasible because both shape the design to maximize total profits. In other words, when a buyer asks a supplier to participate in the development of a new product, the opportunity exists for a design that is both more profitable for the supplier to produce, as well as more beneficial for the client to sell. This results from the innovation dimension: something that is created rather than shared. The innovation dimension makes it possible, though not automatic, that both parties end up in a better situation. As our former colleague Tom

Vollmann would say, it is not just a matter of 'sharing the pie' but, more importantly, of 'growing the pie'.

That is why we speak of **leveraged innovation**. Clients can create the potential for improved innovation by relying on their suppliers' capabilities. By doing so, they are not simply saving on product development support, the primary motivation behind conventional supply chain arrangements. More importantly, they create the conditions for innovation that might provide benefits for all parties involved. Thus, manufacturers have the opportunity to gain more from their suppliers without exerting pressure. Examples exist where the proper mobilization of a supplier's creativity has generated both higher-performing products and lower costs. In such cases, clients get more than they invest – the main purpose of leverage. But leveraged innovation is not automatic. There is a long list of failed cooperative ventures that sought improved product innovation. Leveraged innovation has requirements which, if unmet, lead to failure. Not only should partners' capabilities fit together, the relationship must manage these capabilities effectively. The main purpose of this book is to specify the conditions for successfully leveraging innovation. The following chapter will provide the background for this discussion with an introduction to the concept of 'Early Supplier Involvement', or ESI, the term now applied to this new approach to product innovation.

2 What Is ESI?

By the end of the book, it will have become clear to the reader that ESI is an integrated management philosophy quite at odds with the general perspective of management, at least until very recently, throughout North America and Western Europe. It is not a technique that can be simply grafted on to the existing system. This chapter introduces the concept by first setting the stage with TRW, a classic example of what has come to be seen in the West as the traditional New Product Development (NPD) process, and an outline of the benefits of this traditional process from the perspectives of both buyers and suppliers. In contrast, the ESI NPD process, as used by Diebold, a relative newcomer to the practice, is then presented in detail. The benefits to both buyers and suppliers, illustrated with examples from our research, bring the chapter to its conclusion.

TRW – a Traditional Manufacturer[33]

Motor vehicles and parts constituted North America's first mass production industry and remains a vital sector of the continent's economy. For example, the US industry was valued at US $236 billion in 1992, accounting for 6.5 per cent of all manufacturing employment and 16 per cent of all durable goods shipped. In the same period, automotive components shipments accounted for 4.5 per cent of manufacturing employment and 3.4 per cent of total US shipments. For each $1 spent in the auto parts sector, an additional $2.50 in expenditures and income are generated in the economy as a whole.[34] This segment of the automotive industry is a classic example of one dominated by traditional buyer–supplier relationships. Subsidiaries of the Big Three automotive assemblers – General Motors, Ford, Chrysler – are responsible for 45 per cent of the combined American, Canadian and Mexican automotive components market which was estimated to be worth $90 billion in 1994. The 50 largest independent and predominately US-owned components manufacturers hold another 25 per cent of the market, and approximately 1500 smaller independents the remaining 30 per cent. Owing to the high degree of vertical integration practised by automotive assemblers, sales to those manufacturers often represent a small proportion of the larger independent component companies' total sales. The industry has long had significant cross-border operations and special cross-border trading rights. The establishment of NAFTA in 1994 heralded a period of even freer operations and a more buoyant industry overall. Other trends in this market are the internationalization of operations, including inter-company collaboration, particularly with firms located in Europe.[35]

TRW was founded in 1901 as the Cleveland Cap Screw Company. In 1904, Charles Thompson, a welder with the young firm, developed a new way to assemble automotive engine valves and an important involvement with the automotive industry began to develop. By 1995, TRW was the second largest independent automotive components manufacturer in the NAFTA region with total worldwide sales of US $10 billion and 66 500 employees. North American sales represented 67 per cent of its total revenues. In common with the company's major North American competitors, it had diversified operations along geographic and industrial lines. From an estimated 120 plants in 28 US states and 22 countries, the company generated sales in the auto, space, defence and systems integration businesses. Sales to the auto industry represented US $6.5 billion or 68 per cent of 1995 total sales. Airbags are one of the company's best known products, along with engine and steering parts, seat belts and suspension systems.[36]

TRW's New Product Development Process

Prior to changes during the 1980s, relationships between components manufacturers and their suppliers remained relatively stable for decades. TRW's experience in the early 1990s is a good illustration of how traditional North American components manufacturers have developed new products and integrated suppliers.[37] At the Idea Generation stage, the first of several stages in the new product development process (see Figure 2.1), representatives from TRW's Marketing and Engineering meet for informal discussions until they reach the point where the idea is concrete enough to say 'Let's start designing', and then move to the next step in the process. During Concept Definition, the TRW Design Engineers take

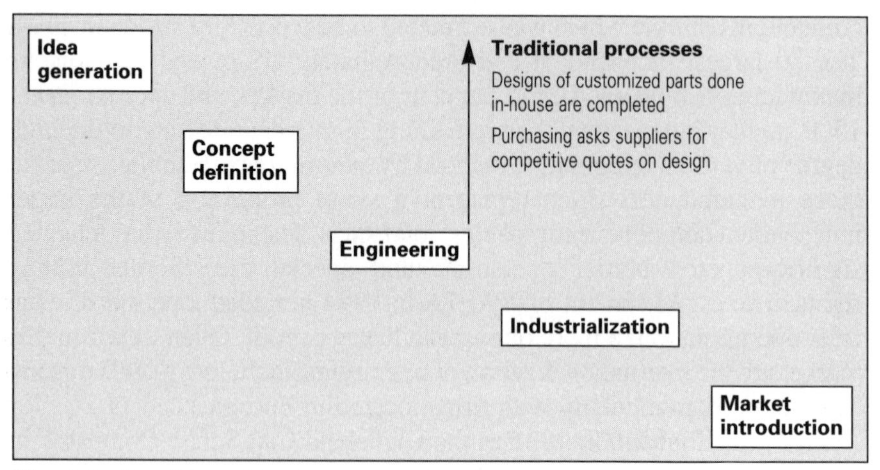

Fig. 2.1 Traditional new product development process

over the idea and begin to work on preliminary designs with their preferred suppliers. TRW Engineers direct the action, develop the drawings, and place the orders for prototype parts. Once the prototype is complete, responsibility turns to the Product Engineering team that works up the final specifications and drawings, which are then passed onto the Manufacturing and Purchasing departments. Next, Manufacturing builds or contracts out the building of the tools, altering the specifications as it finds necessary. Meanwhile, Purchasing lines up suppliers to tender for the business, often the same suppliers that were involved at the Concept Definition stage. Bids are usually requested from three different suppliers, sometimes just two, and on rare occasions only one (for example, in the case of a specialized part that only one supplier has experience producing). Purchasing then selects a supplier to manufacture the part, sometimes in conjunction with Product Engineering and/or Quality Control.

TRW's Relationship with Suppliers

Under the process described above, the assumption made by those responsible for supplier selection is that suppliers who have tendered for business can deliver a good quality part or component, on time. Therefore, TRW does not normally send any of its engineers to the supplier company to inspect the production facilities and evaluate capabilities. As with most traditional buyer–supplier relationships, TRW and its suppliers are not very open with information. TRW offers information about the overall programme on a 'need-to-know' basis. The company's suppliers likewise closely guard information about the costs of labour and materials as well as production processes.

The Traditional New Product Development Process

The so-called 'traditional' new product development (NPD) process, illustrated by TRW in the early 1990s, evolved in conjunction with the rise of mass production. It is equally applicable to many manufacturing players inside and outside the automotive industry. This traditional NPD process is thus widespread throughout those regions in which mass production processes are used most, notably North America and Western Europe.

Mass production evolved in the early 20th Century from the craft production system. Henry Ford, the pioneering US automaker, was perhaps the most influential force behind mass production in manufacturing. As noted by Halberstam:[38]

> When [Ford] began producing the Model T, it took twelve and a half hours to make one car. His dream was to make one car every minute. It took him only twelve years to achieve that goal and five years after that, in 1925, he was making one every ten seconds.

While Ford's contribution to mass production centred primarily on revolutionizing the engineering function, Alfred Sloan, brought in to manage GM in the 1920s, solved many of the problems associated with mass production outside the core engineering function. Specifically, he saw the need for separate functional areas, in particular those of finance and marketing. The division of firms into the full range of functional specialities we know today gave subsequent rise to sequential, disconnected processes across those functional specialities. Functional thinking also led to relations with suppliers being conducted on a similar 'disconnected' basis, essentially the separation of responsibilities behind 'Chinese walls'. Ford and GM's move into Europe early in the century introduced European manufacturers to the mass production process, leading to its widespread adoption, including in the area of buyer–supplier relationships.[39] By the late 1980s, some firms on both continents began moving away from the traditional buyer–supplier relationship as described here. Many, however, did not, and in any case, the fact that this system operated for many decades (and continues to operate) suggests that both parties in these traditional relationships consider the arrangement mutually advantageous.

The Buyer's Perspective

In conjunction with the mass production mentality, buyers perceive the traditional buyer–supplier relationship as offering several benefits, the most important of which is the effect of competition on piece price. Sourcing flexibility and bargaining power are the two other major benefits. Where there are benefits, there are costs, however, and these costs will be analysed in Chapter 5.

Competitive Piece Price

Normally, as in the case of TRW, about three suppliers are asked to tender on the production of a new part.[40] The buyer provides prospective suppliers with final drawings of the part to be produced, the production volume required, and the desired delivery terms. The tools are usually provided and paid for by the buyer. The manufacturer may build the tools itself or contract the building to a tool maker. Sometimes the parts supplier is responsible for the tools. In these cases, the parts supplier signs a separate contract for tool building. Ultimately, whichever of the three parties builds the tools, they are owned by the manufacturer. Suppliers, therefore, usually quote a 'piece price' and delivery terms for the production of the volume of parts specified by the buyer. A single supplier is typically chosen to produce the new part and a purchasing contract is signed between buyer and supplier. If capacity is an issue (for example, if all tendering suppliers are very small and individually lack the capacity to produce a sufficient number of parts within the deadline), a second supplier may be designated. Buyers feel that the competition generated by the tendering process causes the suppliers to quote the lowest piece price they are capable of producing. Often the suppliers and the buyer will be known to each other; that is, the suppliers will have a history of working with the buyer on previous contracts. This history, it is believed, helps to increase the desire of suppliers to win the business[41] and keep the quote as low as possible in order to maintain the relationship.

Sourcing Flexibility

A second feature of the classic tendering process is the flexibility it allows. Buyers usually ask more than one supplier to quote a piece price for the production of a new part. Although only one supplier is usually selected, the tendering process has already offered the buyer a sense of the broader market. If the selected supplier does not work out – for example, proves incapable of producing the required volume of parts within the deadline or is unable to produce parts of sufficiently high quality – the buyer already has the other suppliers involved in the tendering process as backups. Since the buyer owns the tools, they can be moved to one of these two suppliers at the buyer's will. Even if these particular suppliers are not available to produce the parts, with the tools already built, the buyer can go to the free market and look for another supplier with spare production capacity.

Bargaining Power

Aside from the benefits in piece price and sourcing noted above, buyers often perceive a third benefit in the tendering process: that of bargaining power. What traditionally differentiates one supplier from another is price and delivery terms (and, of course, its reputation to live up to the terms it says it can meet). Assuming that suppliers quote only terms that they can actually deliver, then from the perspective of the individual supplier, the company must be selected a sufficient number of times in order to ensure its survival. The buyer commands virtually all bargaining power in the relationship, including, importantly, ownership of the tools.

The Supplier's Perspective

Despite the buyer's overall power in the traditional relationship, suppliers also perceive positive aspects of the tendering process. They typically identify three benefits: controlled costs, information and risk, which compensate for their lack of bargaining power. Again, all benefits have their downside, and these will be discussed in later chapters.

Controlled Costs

Many costs associated with a contract are well controlled by a supplier. Development costs associated with prototype development have generally been assumed by the buyer and so capital expenditures are minimized by default. The buyer is responsible for the purchase of tools, the single most expensive capital item. Overhead, such as electricity, can be calculated into the piece price. Labour costs are usually estimated accurately a year forward, the maximum length of most purchasing contracts.[42] But the controllability of material costs can be less certain and, together with labour, rank as the two biggest expenses for a supplier. In some instances, buyers allow suppliers to 'piggy-back' on their materials contracts, thus permitting suppliers to control that cost as well. In the majority of cases, however, this is not the case and suppliers therefore do need to develop a knowledge of their commodity markets. At the end of the day, barring serious unforseen changes in economic circumstances, the supplier can be reasonably sure of the profit margin on a given purchasing contract.

Controlled Information

The tendering process requires the supplier to provide only the piece price and delivery terms for a specific job. As a result, all information related to supplier costs remains in-house and confidential. The supplier does not have to share actual overhead, labour or materials costs with the buyer. In addition, the supplier does not have to show the buyer around the production facilities, and so information related to production processes is also controlled. Controlled information gives the supplier a crucial small measure of strategic power vis-à-vis the buyer as that buyer compares supplier quotes for a part. A well-managed medium-sized plastic injection moulder, for example, will have lower materials costs than a small 'Mom and Pop' operation in the same business. If both quote a job, the medium-sized firm can bid lower, but make an unspecified larger profit margin at the expense of the buyer.

Controlled Risk

Following these two benefits is a third: the level of risk that a (reputable) supplier carries in the process of completing a purchasing contract. It is low relative to that experienced with ESI. Assuming no significant upheavals in the economic environment, all the supplier needs to do is turn out the necessary number of parts within the required time frame.

Diebold: First Steps Toward Establishing ESI Relationships[44]

The ATM industry is entering its third decade. Both in terms of sales and innovations, the industry is growing following record sales in 1996. That year, the global market, dominated by NCR, Diebold and Fujitsu, comprised 72 157 units shipped, up 15 per cent from 62 457 in 1995. The boom has its origins in (1) increased business from banks located outside North America and Western Europe who now feel competitive pressure to 'modernize' and stay in line with developments in the world's industrial regions, and (2) increased sales from banks in North America looking for more ways to reduce branch transactions, while increasing overall transactions (through, for example, off-site cash-only dispensers and multiple-use machines that dispense stamps, pre-paid phone cards, coupons, and so on).[45]

Diebold is the leader in the US Automated Teller Machine (ATM) business, commanding over 50 per cent of that country's market and holding the number two spot in the world behind NCR. In 1995, the company's worldwide sales amounted to $US 863 million with 80 per cent of those sales in the US market. Founded in 1859, the firm began as – and has remained – a security equipment manufacturer. Safes were the company's original product line.[46] Diebold has kept up with the changing needs of its clients for almost 140 years and ATMs are now the company's leading product. The company is also the US leader for electronic and physical security products (for example, safes, surveillance systems and vaults) and a specialized niche involves security systems for hospitals and universities.[47]

Diebold's Supplier Programme

Despite Diebold's continuing success, by the late 1980s the company felt pressured to streamline its supply base in order to leverage buying practices and remain competitive. Diebold began to establish a company-wide preferred supplier list; that is, to focus efforts on developing a smaller number of active suppliers who could be certified to the firm's quality standards and converted ultimately to ship-to-stock (STS) suppliers. In 1989, the firm had 1250 production suppliers, many of which were third-tier. By 1994, with the initial 'weeding out' process nearly finished, that number had been reduced to 400 and more cuts were foreseen. In conjunction with this process, Diebold initiated and maintains a goal of sharing its quality strategy with suppliers, so all potential STS suppliers receive a copy of Diebold's in-company ship-to-stock manual. The suppliers therefore know exactly what Diebold's quality process is and what they need to do to qualify.

Diebold first experimented with ESI as early as 1990. By 1994, suppliers were being involved early in the engineering phase (see Figure 2.2). When Diebold engineers identified a part (for example, a money cassette for use in a new ATM machine) that could benefit from a supplier's early input on the design of the part's details, they looked through the preferred supplier list and invited suppliers to discuss matters. Initially, they might consider several suppliers for a product while capacity and other basic factors were determined, but they were usually able to narrow it down to one in a short time. The company's ultimate objective

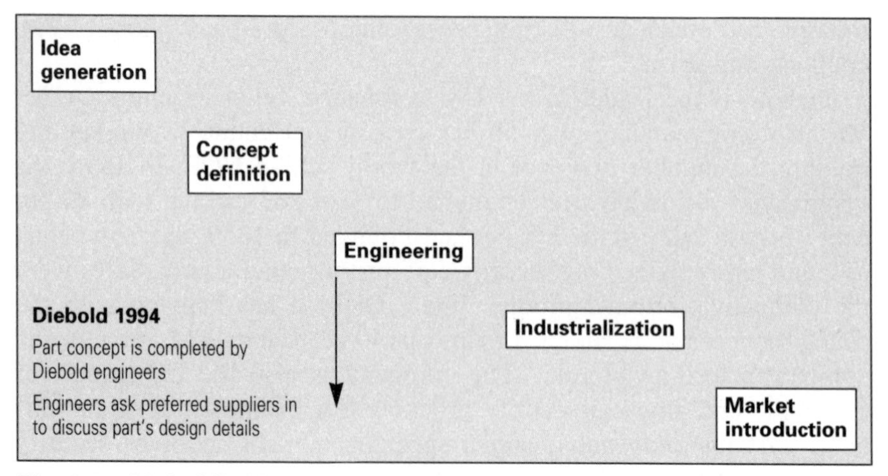

Fig. 2.2 Diebold's New Product Development Process in 1994

was to move to one supplier per 'sub-commodity' (for example, plastic could be broken down into several sub-commodities, an example of which is structural foam).

Once a supplier was selected, representatives were invited to Diebold to participate in setting the target cost and development schedule. The earlier the supplier was brought in the better. Documentation of the decisions taken at this stage, and the reasons behind them, was considered critical. Diebold used a project sheet (completed at the meeting between the company and the supplier) to show what had been agreed upon, including a cost breakdown (material, labour, profit). It was important to understand and fix expected costs up-front. If costs turned out to be higher later on, both Diebold and the supplier could backtrack and understand why (for example, an original material was replaced with a more expensive one). Initially, some suppliers hesitated to give this information but with experience they began to understand that target costs kept people on track. As the project proceeded, the development team needed to deal through buyer-engineer to supplier-engineer on technical details and buyer-purchasing representative to supplier-sales representative on commercial aspects. Additionally, there was a need for periodic reviews with the whole team. At the conclusion of a project, it was also considered essential to sit down with the supplier and review what happened, so that both sides could learn from the experience.

By 1994, Diebold was using suppliers to design the details of most parts, but not to produce the entire design. Co-designs were completed much more quickly and cleanly than designs managed internally or through a design house. The barrier to achieving completely supplier-developed designs was the supplier's capacity to do the design work. This barrier was one of the reasons Diebold sought to further reduce its supply base. With an even smaller supplier base, Diebold could fully support individual suppliers and suppliers would be willing to invest in developing design capabilities knowing they would get steady design work in the future. Ultimately, the company wanted to be able to hand the supplier an idea with an outline of the dimensions and, possibly, materials suggestions, and say, 'We want to work with you, here's our targeted cost and assumptions, do you agree?'. The supplier would then sign a contract, and start work. Each month the team would be brought together to see if the development was on schedule and costs on target, thus leading to a preference for more local suppliers.

Diebold's ESI Challenges

Because both customer and supplier are under pressure to perform, ESI presents the challenge of taking the necessary time to agree on target costs and schedules up-front. A lack of discipline at this stage is especially critical if a change occurs after the process has started (for example, increased costs resulting from supplier changes made on behalf of a customer engineer without previous agreement on target costs and scheduling). Engineers do not naturally stop to ask, for example, how much a particular increased tolerance might cost. Any change in tolerances/materials/other factors will have an impact on schedules and target costs. This issue also arises in traditional buyer–supplier relationships; however, an ESI relationship will only survive in the longer term with openness, coordination and discipline.

A second issue is that a good manufacturer of parts is not necessarily a good designer of parts. This discrepancy in supplier capabilities was discovered during Diebold's reduction of the company's supply base and qualification of STS suppliers, and led to a shift in appropriate suppliers. Now suppliers need to be able to take an electronic data file and build the part. Sometimes Diebold even has parts beginning production before the drawings have dimensions on them. The company learns more about a supplier's capabilities through ESI than by any formal audit, and therefore knows what to do differently on the next project. There is a need to assess the supplier's capabilities up-front on a project-by-project basis. An early example of this was a project that involved a supplier's design input to enhance the facia of an ATM design. Diebold overwhelmed the supplier with too many tools, despite Diebold's long familiarity with the supplier. A supplier quite naturally will not say that its firm cannot handle a customer request when the firm wants the business. Since Diebold works with relatively small volumes and primarily custom-made parts, and so has little choice but to deal with small shops, it is difficult not to overwhelm a supplier on a major redesign.

Diebold's New Relationship with Suppliers

In contrast to typical traditional buyer–supplier relationships, which are characterized by a lack of transparency, Diebold's new supplier relationships have openness as the central tenet. Although inadvertent assumptions can still be made about a supplier's capabilities, establishing an ESI relationship means getting to know a company intimately. As previously outlined, Diebold makes a concentrated effort to assess a supplier's production facilities and production capabilities.

By 1994, after approximately four years of experience with ESI, the process had yielded a 50 per cent reduction in development time, cycle time and speed-to-market. Although ESI may not be the only factor in reducing cycle time, it is certainly the primary one according to Diebold.[48] For this company, the conclusion is obvious, considering the time involved in the traditional 'bid and quote' system.

Perspectives on the ESI New Product Development Process

> ESI is often a matter of survival. Many customers say to us that if they do not capitalize on the resources of their supplier, they will be in trouble.
>
> Gary Deaton, Manager of Marketing and Manufacturing, The Minco Group, Dayton, Ohio

In a world where ESI is increasingly seen as fundamental to a firm's survival, it is important to manage carefully the challenge it poses. A successful transfer of responsibility to suppliers in the area of new product development requires manufacturers to work closely with the supplier. The investment needed to do so is expensive in terms of time for the manufacturer. The company therefore needs to be selective, working closely with fewer, but stronger, suppliers. This selectivity in turn leads to increased dependence on those suppliers. Thus, the first step to a successful ESI relationship is to select the right component or subsystem as the basis both for a new product development and a new supplier relationship. ESI is used most effectively in the design of custom-made parts. The Japanese automotive industry, where ESI originated, simplifies the identification of appropriate parts by segmenting all parts into three categories:

1. Standard parts – for example, nuts and bolts, where no customization is required
2. 'Black Box' parts – for example, Anti-lock Braking Systems, where the development has always been the responsibility of the supplier
3. Custom-made parts – for example, bumpers, which manufacturers used to design with limited or no contribution from their suppliers, and thus where ESI has potential

How the supplier is chosen at the Concept Definition stage differs from firm to firm (Figure 2.3). Some use a design competition in contrast to a traditional price competition described earlier. Others dispense with formal competition altogether and approach a supplier whose capabilities are known to the firm. Still others use an approach that falls somewhere between these two. Compensation for the supplier's design input also varies. Some receive a separate design fee, others work solely on the basis of a potential manufacturing contract, and yet others have the fee for design rolled into the price of the tooling.

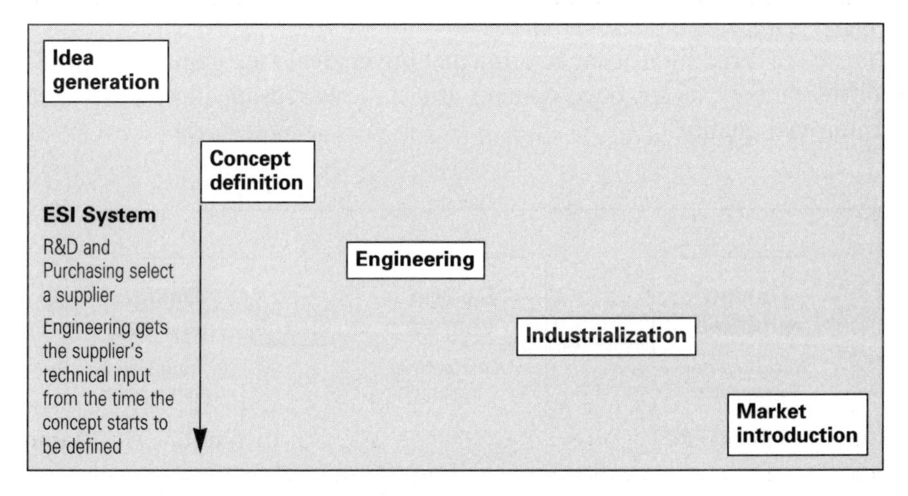

Fig. 2.3 ESI new product development process

ESI is not a 'static state' that a company does or does not practise, but a continuum of increasing involvement of the supplier in the design process which ideally begins at the start of the Concept Definition stage (see Figure 2.3). At the lowest level, 'Design supplied' (see Figure 2.4), the manufacturer takes full responsibility for the product design, and the supplier simply shares information about its equipment and capabilities. A company can then move to the next level, 'Design shared', where the manufacturer still takes full responsibility for the product design, but the supplier provides early feedback on the design, including suggested cost, quality and lead-time improvements. An excellent example of a development handled at this level would be the frame or chassis for the 'Liberty' laser printer co-designed by Lexmark International and its long-time plastics supplier, the Minco Group. Their experience is detailed in later chapters.

Firms can then aim for an even higher level of involvement, 'Design sourced'. Here the supplier takes full responsibility for a system or sub-assembly from concept to manufacture, incorporating one or more parts which the supplier also designs based on an in-depth understanding of the manufacturer's requirements. An example of a development managed at this level is a paper tray subassembly for a photocopier subcontracted by Xerox to a first-tier supplier. That some buyers and their suppliers are experimenting with ESI suggests that both parties not only perceive benefits, but benefits greater than those realized through the traditional tendering process.

The Buyer's Perspective[49]

There are three significant benefits that buyers feel they gain from an ESI development: faster development times, lower production costs and improved quality.

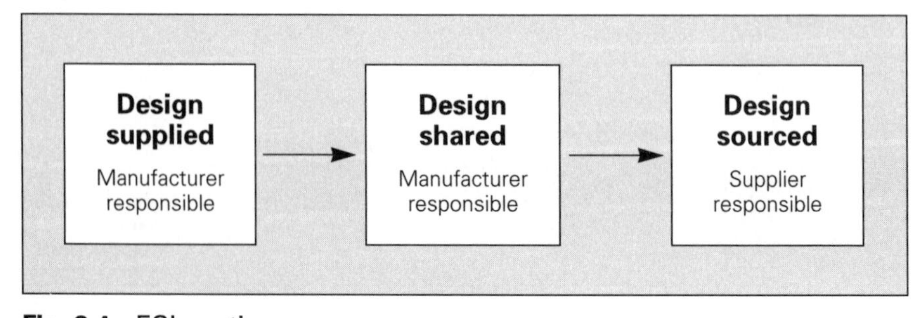

Fig. 2.4 ESI continuum

Faster Development Time

Buyers experience faster development times with ESI projects, despite the huge up-front time investment. Improved development times are largely the result of time savings during the tooling phase. As the tools are built while the design is being developed, time savings come from parallel work and the resulting limitation of rework. On early ESI projects, overall development times can be reduced by 30 to 50 per cent.

> By the early 1990s, Xerox had slashed tooling times by 60 per cent on ESI projects compared with traditional projects. The firm accomplished this tremendous time reduction by working in parallel with suppliers organized into tiers as they were in the automotive industry. Thus, Xerox would contract the paper tray assembly for a new photocopier to a first-tier supplier that, in turn, would contract the paper tray itself to a second-tier supplier. In some instances, even a third tier of suppliers might be involved.

Example 1 Reduced tooling times

Lower Costs

ESI leads to cost savings by simplifying assembly and reducing tooling charges and warranty costs. When a project is viewed 'holistically' from the outset, the overall project design is simpler, and therefore so is the product's assembly. A holistic approach also results in a reduction in tooling charges as the tools, once built, are much less likely to require modification through discovery of errors. Lower error rates lead to additional savings in such areas as warranty costs.

> The Liberty printer development undertaken in 1991 by Lexmark in conjuction with three ESI suppliers, and described in detail further on in this text, resulted in an 80 per cent reduction in warranty costs compared with traditional developments. A holistic approach to product design at the outset led to many fewer problems in mass production and, consequently, a better quality product and significant savings over the life of the product.

Example 2 Reduced warranty costs

Kodak reported that, prior to the introduction of ESI, if $30 million worth of tools were required for a project, then the actual cost would be $30 million plus about twice the magnitude of the programme's estimated cost due to changes, mistakes and misunderstandings. With ESI, the contingency money for tools dropped to less than 10 per cent of the entire programme, in direct opposition to the assumption that ESI might lead to more mistakes through early design commitments. Kodak drove this percentage even lower through further supplier involvement. Electronics suppliers throughout the Pacific Rim provide 50 per cent of the value in a single-use camera. Since cost is a key driver for this product range, Kodak needed to work very closely with those suppliers.

Example 3 Reduced tooling charges

By the early 1990s, Xerox no longer bought individual parts, such as paper trays, for the firm's photocopiers directly from suppliers, but instead contracted entire subassemblies, such as paper-feeding mechanisms. The company managed this process with multi-functional product teams as follows. The supplier would be made a part of this team before any lines were put on paper. A paper-feeding mechanism might be composed of ten plastic parts, yet the Xerox product team would not want to deal with ten different plastics suppliers. Additionally, if those ten plastic parts were the majority of the subassembly, Xerox would have one plastics supplier do the assembly of that component. There was no mark-up on the ten plastic parts, giving the company very good value for money. Xerox also required fewer internal staff to deal with one primary supplier as opposed to ten individual parts suppliers. Complex assemblies were also generally given to local suppliers. If need be, the entire design team could simply get in the car and drive to the first-tier supplier. This way everyone became more productive.

Example 4 Simplified assembly

Improved Quality

Quality improvements are achieved in two ways. The manufacturing process is often simplified, significantly reducing the requirements for rework and emergency engineering changes. ESI also increases the level of innovation, by allowing the suppliers to draw on their own expertise to suggest significant product improvements over the previous generation's design.

Motorola first began to experiment with ESI during the late 1980s when the company realized that it was giving suppliers designs that were not manufacturable. These problematic designs were leading to long cycle times and, consequently, high costs. Motorola's products were becoming less and less competitive in the marketplace. By not understanding supplier processes, the company was essentially paying for the manufacture of scrap. So the company started to do some benchmarking around the issue of manufacturability. At the same time, suppliers to the company were saying, 'Listen to us before you hand us a design.' The result was a very committed and formal move to ESI, leading to a rapid increase in the manufacturability of Motorola products.

Example 5 Simplified manufacturing process

Kodak first began to experiment with ESI in 1988. In the early 1990s, the company launched a new camera called the Cameo. It was designed as a pocket camera, and unlike many pocket cameras, the intention was that the camera should actually fit into the average pocket. This feature, 'pocket-size', was one being demanded by customers, and so Kodak set out to comply. Kodak approached its suppliers early in the concept definition stage to ask for their assistance on a number of features, the aim of which was to fulfil the 'pocket-size' objective. A good example of a feature on which Kodak needed design assistance from its suppliers was the folding viewfinder. A folding viewfinder would help the product development team reach 'pocket-size' in two ways: (1) the overall circumference of the camera would be smaller; and (2) the outside casing would be smoother. The supplier with the right capabilities was able to design the important viewfinder spring. With a range of suppliers involved in the design of other features early on, the overall development cycle time was cut in half to 18 months, and the Cameo was a great commercial success.

Example 6 Increased supplier innovation

The Supplier's Perspective

Although all three benefits that suppliers experience in classical tendering relationships disappear in true ESI relationships, suppliers perceive new benefits, including the opportunity to develop competitive advantages, increase R&D effectiveness, and embrace organizational and personal challenges.

Competitive Advantage

As outlined earlier, the classical tendering process results in few, if any, means by which suppliers may differentiate themselves. Price and delivery terms are the main factors by which buyers select suppliers. An ESI approach allows suppliers to develop and promote unique capabilities for which they can hope to become known throughout their industry. Unique capabilities might be developed around expertise with certain materials, as is the case with plastics. Many products formerly made principally of cast steel parts are now being made entirely, or to a much greater extent, of plastic. There is literally no application that people are not trying to challenge with plastic replacement. Plastic manufacturing requires a different sort of expertise than steel. Once a steel part is cast, it retains its size and shape. Plastic, however, shrinks and warps during the manufacturing process. All plastics do this in different ways and to different degrees. Companies that are switching from steel to plastic manufacture may find that they have limited in-house expertise, and can indeed benefit from cooperating with suppliers who work with plastics every day. Suppliers in turn can develop competitive advantages by specializing in different families of plastic.

In the early 1990s, Nypro, a high-end global plastic injection moulder based in the US, found itself dealing on a regular basis with customers that were used to working in metal, but needed certain parts made in plastic. Plastic conversion allows weight reduction, cost reduction, increased speed to market, and part simplification. Plastic allows the simplification of parts because they can be manufactured in one step, unlike metal parts which require successive steps. On one development programme for a computer manufacturer used to working in metal, Nypro managed to cut the number of parts in half.

R&D Effectiveness

The unique capabilities which suppliers develop to promote themselves vis-à-vis competitors are centred on R&D expertise, and, if well managed, lead to increased effectiveness of the supplier's (now more specialized) R&D function. A more effective R&D capability, carefully chosen, should lead, in turn, not only to an increase in customer demand for the supplier's services, but should also contribute to increased profit margins. The experience of Sonceboz illustrates this potential.

Sonceboz is a small Swiss supplier specializing in electric and electronic motors. During the 1980s, the company became increasingly specialized in the development of small motors. The company's decision to focus its R&D efforts in this direction led to the company's involvement in a major advanced research programme. In the mid-1980s, BMW started to consider a new development which included a specialized small motor application. Two years into the development programme, Siemens was brought in to assist with certain aspects of the development, and a further two years later, Siemens brought Sonceboz on board. At the request of BMW, Siemens hid the name of the company behind the development from Sonceboz. In 1990, BMW decided to move from advanced design to new product development. It was at this point that Sonceboz learned from Siemens that the sponsor company was BMW. In 1994, mass production of the new product began, including Sonceboz's small motor. All Sonceboz's contribution legally belonged to Sonceboz. The company was protected for three years from mass production against BMW going to Bosch, Magneti Marelli, or any automotive supplier for the same motor. Sonceboz argued that, because the specifications of the programme were so precise, the know-how that the company brought to the development had to remain its property. Sonceboz would not have got involved in the programme on any other basis. A major benefit of the company's involvement was increased R&D effectiveness in its chosen area of expertise.

Increased Challenge

Following both of the above-noted benefits, the third benefit suppliers feel they experience as the result of involvement in ESI NPD projects is increased personal and organizational challenge. Suppliers that have shifted their participation from involvement at the end of the industrialization phase of the new product development process, to deep involvement beginning at the concept definition stage are, as described in Diebold's case, dealing with a more organizationally complex cross-boundary relationship. R&D engineers also need to push the boundaries of their knowledge to design and build the tools in conjunction with the design of a part.

For example, Nypro, the high-end plastic injection moulder already mentioned, is a company always looking for challenges. As Nypro is a job shop (in the company's own words, 'the lowest form of manufacturing species'), it lives and dies by the successes and failures of its customers.

The higher up on the 'food chain' Nypro is, the lower the level of price competition, the more the company can add value on development projects, and the better it can survive the ups and downs of the industry and the economy in general. So Nypro always tells the customer that it can make the part requested, but asks whether the customer would like Nypro to do more than just manufacture the part. In 1994, Nypro was already doing subassembly (decoration, joining, bonding, coding, combining) on Motorola flip-phones. That same year, Hewlett-Packard asked Nypro to take over its plastics technical centre. This move took HP out of the design phase and brought Nypro one notch up in the food chain.

Conclusion

By the late 1980s, a number of firms on both continents began moving away from the traditional buyer–supplier relationship described here. Many, however, did not, and in any case, the fact that the system operated for many decades (and continues to operate) suggests that both parties in these traditional relationships did perceive themselves to gain from the arrangement. In conjunction with the mass production mentality, buyers perceive the traditional buyer–supplier relationship as offering several benefits, the most important of which is the effect of competition on piece price. Sourcing flexibility and bargaining power are the two other major benefits. Despite the buyer's overall position of power in the traditional relationship, suppliers, too, have their positive perspective of the tendering process. They typically identify three benefits, controlled costs, information and risk, which compensate for their lack of bargaining power.

That some buyers and their suppliers are experimenting with ESI suggests that both parties not only perceive benefits, but benefits greater than those achieved with the traditional tendering process. There are three significant benefits that buyers feel they gain from an ESI development: faster development times, lower production costs and improved quality. Although all three benefits that suppliers experience in classical tendering relationships disappear in true ESI relationships, suppliers perceive new benefits, including the opportunity to develop competitive advantages, increase R&D effectiveness, and enjoy greater organizational and personal challenges.

3 The Origins, Diffusion and Evolution of ESI

Rooted in the increasing socialization of the innovation process, ESI originated primarily in the Japanese automotive sector in the late 1940s where industry conditions were catalytic. The practice later spread to the US automotive industry via Japanese transplants in the early 1980s, and began to diffuse among other regions and industries beginning in the late 1980s. This chapter explains this process through five case histories, beginning with Toyota, the company where ESI originated. The story then moves to the US when Honda, one of the first Japanese transplants, adapted ESI into the North American context. American companies observed the activities of these transplants and tried implementing the concept themselves, Chrysler being the first US automaker to do so. Early non-automotive manufacturing companies to adopt the concept include Motorola. Later the practice moved full circle back to Japan and very sophisticated use by some of that country's non-automotive manufacturers. Concluding the chapter, the history of Fuji Xerox illustrates the latest stage in the story of ESI.

Toyota: Necessity is the Mother of Invention[50]

ESI is best understood by beginning with the post-war restoration of Japan's automotive industry. Although Japanese automobile manufacturing began in the 1930s, events following the Second World War created fundamental changes throughout the country's economy, including the nascent automotive industry. Womack *et al.*[51] summarize the situation at that time as follows:

- The domestic market was tiny and demanded a wide range of vehicles.

- The native Japanese workforce was no longer willing to be treated as a variable cost or as interchangeable parts.
- The war-ravaged Japanese economy was starved of capital and foreign exchange, meaning that massive purchases of the latest Western production technology were quite impossible.
- The outside world was full of established motor vehicle producers who were anxious to initiate operations in Japan and ready to defend their markets against Japanese exports.

It was for these reasons that the country's industrial development generally, and the automotive industry specifically, followed a different path from that of North America and Western Europe.[52] Casual observers of Japan often attribute its success in world markets to the strong role the Japanese government played and continues to play in supporting and protecting the country's businesses. Others see Japan's very competitive internal market as the key to the country's success. Friedman[53] refers to these two perspectives as the 'bureaucratic control' and 'market regulation' theses. In reality, both views are over-simplified. The following example is an illustration of the distortion that either argument alone fosters.

Early in the post-war era, the Japanese Ministry of International Trade and Industry (MITI) imposed a range of restrictions on the auto industry. At one point, however, MITI attempted to force a merger of the 12 existing automotive firms into two or three larger firms, as in Detroit, each of which would specialize in one consumer product segment. The thinking was that this reorganization would allow the industry to benefit from economies of scale, compete with the Americans, and yet, across companies, still satisfy consumer demand. Not only did the 12 refuse to merge, but Honda, a motorcycle manufacturer and therefore ordered by MITI to not produce cars, chose to defy government orders and is now considered by many the most progressive of the Japanese automakers.[54] As testament to this view, the fourth-generation Honda Accord, designed and manufactured for worldwide sales in Marysville, Ohio, was the best selling American-built car from its launch in 1990 through 1992, the first time this honour has not been held by Ford or GM since the launch of the Model T some 80 years earlier. The bureaucratic control and market regulation theses not only distort reality, but overlook a further important aspect of the country's economy: the supplier sector. In contrast to the role played by suppliers in the development of the US economy, Friedman[55] argues that:

> Japanese manufacturing growth resulted ... from the dramatic expansion of smaller producers throughout the nation's economy. Special

> circumstances [especially the lack of capital and foreign exchange noted by Womack et al] in Japan [forced] smaller-scale producers to implement more flexible manufacturing strategies than those which were possible for firms that pursued mass production alone. The result was to enhance the ability of Japanese manufacturers to adopt extensive, continuous product changes more easily than producers in other countries. This flexible manufacturing and ability to modify products was an even more important factor in Japanese expansion than the efficiency gains emphasized by other observers. Indeed, if the Japanese had in fact organized their economy as most popular explanations suggest – reducing costs by coordinating, integrating, or consolidating production – the country's rapid expansion might not have occurred at all.

In the context of the constraints outlined by Womack *et al.*, the flexibility that suppliers throughout Japanese industry were forced to develop created a strong impetus for what are commonly known as 'Japanese management practices'. ESI is but the latest in a long line of such practices including Kanban and Kaizen. In order to understand fully the practice of ESI as a management philosophy rather than just a technique, it is helpful to follow the development of Toyota, the company that is possibly the most Japanese of automakers, and arguably the originator of ESI as a systematic practice.

Toyota's Early Years

Pre-War

Kiichiro Toyoda's interest in the automotive industry began formally in 1929 with a visit to Ford. Kiichiro was a member of the family that ran Toyoda Automatic Loom, and, on his return from the US, he started to experiment with the manufacture of automobile prototypes during idle time on the loom production line. Successes at this stage led to company approval to establish an automobile division, and later a separate automobile company. With the passage of the Automobile Manufacturing Law in 1936, Ford and GM were virtually ousted from the country, clearing the market for the as-yet-to-be-established Japanese firms. The Toyota Motor Company was founded in 1937.[56] From the company's founding through to the late 1940s, the young business struggled to survive. Instead of being allowed to concentrate on the production of passenger cars, with the outbreak of war in Europe in 1939, the government forced the firm to manufacture trucks as part of the country's war preparation effort.

Post-War

All automobile development work ended with the outbreak of war. In the first 13 years of the company's life, Toyota had managed to produce a total of 2685 passenger cars employing largely craft production methods, whereas by 1950 the Ford Baton Rouge plant was turning out 7000 cars every day with mass production assembly lines. The use of 'backward' production methods, however, was not the only obstacle the company faced in 1949.

Owing to economic measures imposed by the occupying American government after the war, Toyota's sales collapsed, and the company was forced to lay off a quarter of its workforce. Efforts to reduce the unionized workforce resulted in a strike that ended only with the resignation of Kiichiro and his acceptance of personal responsibility for the company's failure. Management gave to the remaining workforce two guarantees, which have had a significant influence on Japanese industry: (1) lifetime employment, and (2) pay graded by seniority, rather than by job function, with bonuses tied to company profitability.

In 1950 Kiichiro's nephew, Eiji Toyoda, a talented engineer in his own right, made a pilgrimage to Detroit. He was impressed with what he observed, but also thought he saw ways to improve the system. At the time (and until the 1980s), the average American car body remained in production for ten years. To maintain low costs, the American automakers focused on keeping production runs as long as possible and die changes to a minimum. Each changeover took about a day to complete during which time production on the line came to a halt. With each car composed of 300 steel parts, the potential for time lost through modifications was significant, hence the low variability in product offerings. Thus, as many parts as possible were stamped on dedicated lines.

For the reasons put forward by Womack *et al.*, as mentioned earlier, it proved not only impossible for Eiji to improve on Ford's system of mass production, but even to implement mass production at Toyota. The company was going to have to find a way to stamp all parts from just a few press lines. With the Americans so experienced at the low-cost, low-feature game, Eiji felt that the only way for his company to compete in the world market was by offering quick feature changes. Quick modifications meant quick die changes to allow for profitable short production runs. Eiji and Taiichi Ohno, the company's head of production engineering, began to work on their own new system, elements of which were Taiichi's now world-famous just-in-time concept as well as American quality techniques advocated by Juran and Deming. The overall system that they developed came to be called lean production, in contrast to Ford's system of mass production. With a concerted effort, by the late 1950s the company had managed to reduce the time needed for a die change from one day to three minutes. Further, the institution of on-the-line problem solving meant that, although the line stopped almost all the time in 1950, by the late 1970s the line almost never stopped. It was to prove fortuitous for Toyota that the company developed the lean production system when consumer demands for product differentiation and product reliability began to make themselves felt.

Toyota's Early New Product Development Process

Pre-War

In the early 1930s, the very earliest days of automobile manufacture at Toyoda Machine Loom company, the new product development process consisted of buying a Chevrolet, reverse-engineering it, sketching the individual parts, and estimating the types and quantities of materials used. Most manufacturing at this time was carried out using the flexible and general tools of the craft production method, and so product development was an integral part of the production process. The individual modifications that went into each vehicle meant that parts used for one vehicle were often not serviceable for use on another. In addition, the more complex parts such as gears were initially beyond the scope of the young company, and some American-made mass-produced parts were integrated into the system. Although each vehicle produced was unique, satisfying the individual needs of each purchaser, the process was long and expensive, and production volumes very low. The result was a kind of 'American-copy' craft-produced vehicle.

Post-War

The American car companies focused from very early in their histories on the separation of functions and, within those functions, on the development of sub-functional specialities. Thus, within design engineering, one found brake design specialists, steering design specialists, engine design specialists, and so on. So too, within manufacturing, one found brake manufacturing specialists, steering manufacturing specialists, engine manufacturing specialists, and so on. Each and every part had its pair of specialists who worked independently of each other. Womack *et al.* cite a wonderful example of just how narrow this specialism went:

> Professor Kim Clark of the Harvard Business School reports finding an engineer in a mass-production auto company who had spent his whole career designing auto door locks. He was not an expert on how to make door locks, however; that was the job of the door-lock manufacturing engineer. The door-lock design engineer simply knew how they should look and work if made correctly.
>
> Functional divisions combined with extreme specialism meant that truly integrated product development teams did not really exist. Teams would indeed be created, but the team leader had little authority, and was really able to act as little more than a coordinator. Individuals on the team were really only motivated to satisfy the hierarchy in their functional sub-specialties. [T]he traditional product development process often led to compromises in the overall design.

As Eiji and Taiichi started to develop the system that was to become lean production, they reasoned that there was much to be gained from a truly integrated product development team where the loyalty of the individual team member was to the product development team rather than to functional divisions and individual sub-specialities. And so, all those necessary to the development of a new product – design and manufacturing engineers, marketing and purchasing personnel, and later suppliers and customers – became first and foremost members of that team and were rewarded on a team basis. The first step Toyota took in creating such teams was the training of all production workers not only in a variety of basic skills, such as machine repair and materials procurement, but in a variety of job functions. This strong team approach immediately eliminated the Detroit-style focus on sequential product development, and created in its place the opportunity to develop concurrent processes.

Toyota's Relationships with Suppliers

Pre-War

In the first years of automotive manufacture in Japan, an automotive supplier industry as such did not yet exist. In 1930, the number of suppliers that did automotive work was 30. In 1931, encouraged by changes in Japanese government legislation, American car manufacturers began to subcontract the manufacture of certain parts to Japanese suppliers. Production by indigenous Japanese firms was also on the increase. By 1938, the number of suppliers had more than quadrupled to 136. At that time, the government recognized the importance of the industry by imposing registration requirements and the like on players in the industry. Thus began the growth of the industry which was to become so important in the post-war success of Japan.

Kiichiro's initial efforts at subcontracting started with trying to locate firms capable of producing copies of the parts from the reverse-engineered Chevrolet, and later from individual parts purchased from abroad. The technological base of the country was very weak, and so the quality of the first parts was very low. Consequently, the company was often dissatisfied with what its subcontractors produced. The number of defective parts was very high, especially in electrical components. Toyota could not vertically integrate these suppliers into the company as the American firms had done in an effort to control quality, because, like many other Japanese manufacturers, the company lacked the capital to do so. Relations between manufacturer and supplier were very poor.

In 1936, when Toyota's output totalled only 200 vehicles (cars and trucks), 51 per cent of the manufacturing cost per vehicle was purchased. In the late 1930s, the Purchasing function began to be handled by an independent department, rather than by a sub-department of Manufacturing as was typically the case in the US, according its personnel more status relative to their American counterparts. It was also at this time that Toyota began to create supplier groups in order to address the quality and technology issues. By 1939, when a total of 12 000 cars and trucks were produced, between 70 and 80 per cent of the manufacturing cost was purchased.

In 1940, Toyota formally classified its suppliers into three categories. Ordinary outsourcing suppliers, the first category, made parts using general purpose equipment. All incoming parts from these suppliers were inspected. The company's policy was to switch among these firms as it deemed necessary. Special outsourcing suppliers, the second group, were those that required specialized equipment and technical assistance from Toyota. As a result, Toyota developed close financial ties with them. These suppliers were as involved in the development of prototype parts as their capabilities allowed. Final parts were inspected at the supplier's premises. The final category, specialist factory outsourcing suppliers, also produced parts that required specialized equipment, but had not as yet developed close capital and financial links with the assembler. These suppliers were earmarked for closer relationships in the future. As with the second group, parts inspection was carried out on the supplier's premises.

At least as early as 1940 Toyota recognized the need to establish and maintain close long-term relationships with some of its suppliers in order to provide them with technical assistance and specialized equipment, and in return benefit from the suppliers' prototyping capabilities and possibly, although this point was not explicitly made, their design capabilities.

Post-War

Although the seeds of the current supplier system were sown in the pre-war era, Fujimoto[58] tells us that:

> In the early 1950s, the Japanese automobile supplier system was very different from what we see today. Many of the basic patterns of the so called Japanese supplier system, including long-term relations, multi-layer hierarchies, 'Alps' structure, Just-in-Time delivery, subassembly of components by first-tier suppliers, involvement of first-tier suppliers in product development, competition by long-term capabilities, close operational control and assistance by the automakers, etc., were gradually formed in the 1950s to 1970s. The high growth of production volume and proliferation of models during the 1960s facilitated the formation of multi-layer hierarchy of control and assistance.

Kiichiro stated shortly after the war ended:

> I want to change the parts manufacturing policy drastically. In the past, for various reasons, Toyota made many parts in-house and thus could not concentrate on parts procurement. From now on, we would grow specialization of our suppliers, have them do research in their specialty, and nurture their capability as specialist factories. We will ask such specialist manufacturers to make our specialist parts.[59]

The company began the introduction of lean production techniques into the supply chain, in what turned out to be a 20-year process, by creating tiers of suppliers organized by type of responsibility. First-tier suppliers were responsible for a complete subassembly such as a steering system. These suppliers developed relationships with second-tier suppliers who developed the individual parts that made up the subassembly. In some instances, even a third tier of suppliers might be needed, depending on the complexity of the subassembly. In-house parts production was spun off into quasi-independent first-tier firms in which Toyota maintained a small equity stake. The company also purchased shares in the independent companies it wished to develop as first-tier suppliers. Toyota further encouraged first-tier suppliers to share personnel as necessary and to buy shares in each other's firms, so that their destinies were all linked. Winning business from rival automotive manufacturers and firms in other industries was encouraged for the extra profits it would bring not only to the suppliers, but also to Toyota.

First-tier suppliers needed expertise in product design and project management whereas second-tier suppliers were required to be skilled in manufacturing and process engineering. Toyota pushed the development of supplier capabilities by providing first-tier suppliers with a minimum of details, perhaps only performance specifications, and encouraging interaction not only between themselves and the suppliers, but also between suppliers in order to solve problems. The institution of this method of working could be seen as the start of an ESI-style new product development (NPD) process.

Nippondenso, now the largest electrical and electronic systems and engine computer manufacturer in the world, is an important, albeit somewhat special, illustration of the supplier structure Toyota created. Originally an electric parts factory that was losing money, it was spun off from Toyota in 1949 as a quasi-independent company. Toyota retained a small equity stake in its new supplier. All of the company's electric components engineers were transferred to Nippondenso, thus providing the new company with immediate design and engineering capabilities. These engineers were already developing drawings based only on sketches of US parts, and so, as members of a first-tier supplier firm, were well on their way to the first systematic 'black box' parts system,[60 61] a process which today might well be considered 'beyond ESI'.

Honda: Making the Practice of ESI Its Own[62]

The Evolution of ESI in the Japanese Automotive Industry

According to Toyota, the company took 20 years to implement lean production throughout its supply chain. The company's production system, including product development methods and ESI, were copied by much of the Japanese automotive industry.[63] Fujimoto reports in detail on the diffusion of black box practices:

- The peak period of the institutionalization of the approved drawing system,[64] as well as actual diffusion of the system within each supplier, was the late 1960s. The next wave came in the 1980s.

- The peak time for the start of informal requests [for approved drawings] from the automakers was the early 1960s. The pattern of the informal requests tended to precede that of formal institutionalization.
- Many of the first-tier suppliers started to regularly hire college graduates for engineering jobs during the 1960s. Establishment of formal engineering sections or divisions in and outside the factories tended to precede full-scale activities in product engineering.
- The suppliers started to make engineering proposals and conduct VA-VE[65] activities mostly after the 1970s.

Womack *et al.* argue, nonetheless, that lean production in its fullest sense has yet to appear anywhere in the world, and that not even the Japanese have fully grasped what truly global lean production means. Honda is the company that comes closest to a lean production organization and hence to ESI on a global scale.

The company was founded in 1948, made its first car prototypes in the early 1960s, and started car mass production in 1967. Unlike the rest of the Japanese assemblers, the company has never developed a kyoryokukai (supplier grouping). As of 1993, Honda Japan had 32 affiliated suppliers and 276 independent suppliers. Most of these independent suppliers have substantive non-automotive businesses, in contrast to the typical kyoryokukai supplier. Unlike manufacturers that have kyoryokukai suppliers, Honda treats all 276 of its independent suppliers equally.

Moving offshore for the typical Japanese automotive manufacturer means moving away from its kyoryokukai supplier base. The fact that Honda has never operated with kyoryokukai suppliers may be one of the reasons it has been the leader among Japanese assemblers in developing global operations. The company began US operations in 1979 with the production of motorcycles in Marysville, Ohio. Car production began on the same site in 1982. By 1992, Honda US was exporting cars to eight countries, including Japan.[66]

The New Product Development Process at Honda US

Three to four years before a new car model is to reach mass production, a team of several top-level Honda directors, representing Sales/Marketing, Product Engineering, Manufacturing and Manufacturing Engineering, is selected to carry out early conceptual work about how the customer will interact with the product (how he will feel when he's driving the car; what he will be thinking; what sensations he will be experiencing, and so on). As of the mid-1990s, all top-level directors were Japanese nationals based in Japan.

Following this brief period of conceptual work, the team visits the Product Engineering area which is comprised of 6000 Honda and supplier engineers. Most of these people are located in Japan, although there are now some in the US.[67] Product Engineering is subdivided into component groups like lighting, steering, and so on. The team makes a shopping tour of the area to see what features are available in lighting, for example, for the new model. The engineers in this product group, together with their suppliers, work their entire careers solely on lighting, developing a system for the 00 Civic (which is just about complete three years ahead of mass production), then one for the 02 Accord (which is at a more conceptual stage), and so on into the future.

Once the top-level team has selected the general features for the new model, a product development team is formed representing, as it does in Toyota, design, manufacturing engineering, purchasing and marketing personnel, suppliers and customers. Honda uses a matrix structure which means that individuals assigned to the product development team come from functional area 'homes'; however, once assigned to the team, they remain with the team for the life of the development. Only once the development is complete do they return to their functional areas.

Honda US: Relationship with Suppliers

For most products in the product engineering group, it is the suppliers that are in charge of the development. The suppliers have people working full-time in the Honda product engineering facility. They also have their own R&D facility nearby. The suppliers develop the concepts, sometimes jointly with the Honda product engineers. The Honda side keeps the suppliers appraised as to proposed new model introductions (for example, a new pick-up truck to be introduced 'X' years hence). Since the early 1990s, this system is starting to be replicated in the US.

Honda US developed the first Accord Wagon, produced in 1990 (that is, the 1991 model). The company designed the load floor behind the rear seat with a supplier in Japan. When Honda wanted to renew the model in 1994, US Manufacturing was involved in selecting a US supplier for the new load floor. First, a purchasing representative and an engineer benchmarked the competitors' products. They then picked four different technologies, found the best suppliers in each of those technologies, and gave them general concepts from which to work (what the appearance and dimensions of the product should be; how the customer should interact with the product, and so on). Honda ended up with six suppliers and four design concepts. These six suppliers had free rein to submit their own concepts.

From the available suppliers and designers, Honda chose one technology and one supplier, Rockwell International. Included in Rockwell's proposal was an 'image' of what the price could be. The proposal was detailed enough to develop a price estimate in conjunction with Rockwell's previous experience developing a similar product for another assembler. Then Honda sat down with Rockwell, and discussed how best to manufacture the load floor. What Honda had from the supplier at this point was a concept with some specifications, especially related to crash-testing. Then Honda engineers at the US R&D facility did the detailed drawing together with the supplier who looked at better ways to manufacture the part. As the project moved further into development, the supplier took more responsibility.

Honda did not select a second source for the new load floor. In general, Honda believes that it is more important to have a backup plan than a second source. Rockwell had several plants that could manufacture the part, and so the supplier worked out how production could be shifted if necessary. Rockwell also had a proven track record with Honda, and had developed a similar product for a competitor.

Honda does not seek a second source to verify the price a supplier is quoting. The assembler had the six original quotes along with sufficient detail from Rockwell to have a good sense of an appropriate price. Honda works so closely with suppliers that a firm quote is not too important. Honda operates with the understanding that the company and the supplier will work together to develop the best processes, and pricing will be worked out concurrently within that process.

Honda has a goal of 60 per cent competition on parts, which means parallel sourcing for 60 per cent of product categories. For example, one supplier makes Accord door-liners and another Civic door-liners, but the Accord door-liner maker could make Civic doors, or vice versa, if it became necessary. The company almost never uses true double sourcing although an exception might be made for capacity reasons.

Honda has a programme called B/P (B/P can stand for 'Best Practice', 'Best Price', or 'Best Process'), the aim of which is to work with suppliers to improve efficiency through all phases of development to mass production. The programme is divided into two stages. The first is called 'soft' B/P, and is used with a new supplier, with that supplier's approval, once the supplier begins mass production of a part for Honda. Honda has conducted soft B/P all over North America with its suppliers. It is a systematic approach to examining manufacturing processes with the aim of balancing the line and improving quality. Two Honda engineers move to the supplier firm full-time for a period of three months. Together with two supplier engineers, they choose a line to use as a model line, and B/P begins. The aim is to train the two supplier engineers and the operators on the model line to the point that they will be able to transfer learning throughout the plant. Both Honda and the supplier benefit from the lower costs resulting from this transfer of knowledge.

After choosing a model line, the first step is usually to clean up the line. For example, at one supplier where soft B/P was carried out, the die-cast machines had been running for years without having been cleaned, and were covered with grime. The team of Honda engineers, supplier engineers and line operators spent the first several days cleaning, putting in new lighting, putting down comfortable mats for the operators to stand on, and so on.

Once the clean-up phase is complete, the Honda engineers take the supplier model line operators into a room, ask them how long they have been working for the supplier, and ask for suggestions on improving work processes. When the employees see their ideas in action a day or two after they suggested them, Honda believes they will take greater pride in their work. At the end of three months, there is an evaluation process which is attended by the Vice-President of Manufacturing at Honda US and the President of the supplier company. This phase is called soft B/P, because it is conducted once a supplier is in mass production, and so little or no capital investment is involved. The evaluation process serves to underline the possibilities for further savings for both Honda and the supplier by implementing a similar process, 'hard' B/P, before mass production begins.

Hard B/P needs to be implemented as early as possible, two to four years before mass production. The ideal time is at the concept stage, once Honda and the supplier have decided what the product will look like. It is at this point that capital investment can be most affected by the process. Hard B/P works hand-in-hand with ESI. To illustrate, an owner-operated firm in the stamping business was asked to quote on a detailed drawing (done together by Honda and the supplier's engineers) based on the supplier's normal processes. The resulting price can be thought of as an index of 100 on a scale of 0 to 100. Honda then asked the supplier to apply the lessons they had learned on the model line and requote. The supplier came back a month later with an index of 87. Honda responded with several suggestions and the offer to allow the supplier to buy the necessary raw materials off Honda's purchasing contracts at a price equivalent to Honda's cost. The supplier came back two or three weeks later with an index of 67. Honda pushed the supplier one last time, saying, 'Go back and give it a last best effort, be creative, put it on a different machine', and the supplier came back with a 57. The supplier has now gone through this process several times for several different parts, and, while the process has not always been as dramatic as in the example outlined here, it has always been successful.

Hard B/P is done with a supplier that has won the business to produce a part. It is not a competition to select a supplier. If a supplier is capable, then he generally knows a lot more about his technology, processes and machines than Honda does. Hard B/P is also the way Honda ensures it is getting the best price and of emphasizing to a supplier that if that supplier is interested in improvement, offering good quality at a good price, the company will get Honda's business in the future. The supplier feels more secure and the company develops expertise in a particular product. Honda expects its suppliers to be experts, which is why company managers meet with the model line operators in soft B/P: Honda wants to tap into their expertise. It is Honda's position that, if they work with a supplier to improve productivity and efficiency, then that supplier will be the most competitive in that industry. According to Lamming, one supplier described the B/P process as 'the agony and the ecstasy!'.[68]

Honda Japan has been using the B/P system for the past 15 years. At Honda US, there is still a 'mix and match' process which means that not all of Honda's suppliers have completed both stages of the B/P process. Nonetheless, there is nothing strictly 'traditional' about any of Honda US's relationships with its suppliers. 'Mix and match' means taking the best of the Japanese and American cultures, and not just blindly adopting the Japanese way.

Honda believes it does not really make sense to undertake hard B/P without ESI. It is in the combination of these two techniques that the real gains lie. Few American suppliers have the R&D capability as yet. A good company could probably get there in four to six years, while an average company would take ten years or more. It is not just a question of people, but also of capital investment, CAD systems, learning how to use the technology, and so on. Even so, the first design or two will be difficult. Honda has a guest engineer programme to help with the development process. As R&D capabilities mature in North America, it will become more common to assign supplier engineers full-time to Honda, as has been the case in Japan for many years. Honda also offers a series of training courses for all its associates, suppliers and employees.

Trust from American suppliers was difficult for Honda to obtain initially. But over time most its suppliers have begun to understand what Honda is doing with the data the company is collecting, along with the long-term commitment Honda makes to its partnerships. Suppliers say they see Honda people in their facilities three times more than all their other customers combined. At Honda, there are 800 people – 300 in purchasing, 200 in quality, 300 in product and manufacturing engineering – dedicated to working with 300 suppliers: a ratio of 3 to 1. While other auto companies may have more people, their ratios are lower. Honda's approach to ESI is building trust through interaction.

Chrysler: Necessity is the Mother of Invention Part II

The North American Automotive Industry in the 1980s

In the early 1980s, the American automotive manufacturers began to feel real pressure from the Japanese and their lean production system. It was at this time that the first Japanese auto 'transplants' began operations in the US. Honda began car production at its Marysville plant in 1982, and the NUMMI joint venture between GM and Toyota was formalized in 1984. Honda and Toyota could now compete head-to-head with Detroit's Big Three in the American marketplace. Although the Japanese share of the North American market was only 24 per cent by 1990, the position of the Honda Accord as the best selling car for several years in the early 1990s served to underline the ever-present threat these new entrants to the market still posed. Detroit's position as the worldwide centre of the automotive industry was for the first time in real danger. Chrysler, as the smallest of the Big Three, and the only one without significant overseas operations to cushion against domestic economic shocks, was the most immediately vulnerable.

Chrysler

Historically, Chrysler has been the least integrated of the American Assemblers. Abernathy attributes some of the success at Chrysler to its lower level of integration:[69]

> Because Chrysler produced fewer of its own components, it was less constrained in adopting advanced innovative components. Thus

> Chrysler could seek competitive advantages through flexibility in product engineering and in styling. Chrysler pioneered high-compression engines in 1925, frame designs permitting low centre of gravity in the 1930s and the experimental introduction of disc brakes in 1949, power steering in 1951 and the alternator in 1960.

Nevertheless, at the start of the 1980s, the assembler found itself on the verge of bankruptcy. Lee Iacocca was brought in to try to turn the company around. This he achieved, challenging the company to produce innovative products like the Reliant K-Car. Having initially pulled the company out of trouble, he pushed it back into the red several years later with diversifications into the aerospace and defence industries, two areas where the Japanese were not competitors. The company's financial losses began to escalate. In 1990, sales dropped 17 per cent against an industry average drop of 5 per cent, and the company slipped to fifth place in North American sales – behind not only GM and Ford, but behind Honda and Toyota also. By 1991, there were rumours that the company would have to be taken over by another assembler in order to survive. During this troubled period, Iacocca managed to persuade the US government to guarantee bank loans, while the recently acquired businesses were spun off and the company reorganized. By 1993, Chrysler had become the world's most efficient automaker, earning an average of $828 on every vehicle produced. The company's transformation is widely regarded as one of the biggest turnarounds in American industrial history.

Chrysler's New Product Development Process

In 1989, Chrysler moved away from the traditional product development process used by GM and Ford to concurrent engineering and a team approach for the development of its limited edition Dodge Viper sports car. This was to be the 'test-bed' for further developments. The company aimed to relaunch all models across its entire product range by 1995 at an expected cost of $16.6 billion.

By 1991, with the success of the Viper behind the company, it moved to a company-wide platform team approach, similar to that used by Honda, to manage new product development from the idea generation stage through to final assembly and mass production. Four teams were created, one each for small cars, large cars, Jeeps and trucks, and minivans. Each platform is headed by a general manager who accords other functions such as Design and Purchasing a status equal to that of Engineering.

With the LH car project, which led to the 1992 relaunch of the company's line of mid- and large-sized executive cars including the Chrysler Concorde, Dodge Intrepid and Eagle Vision, product development time was cut by one full year to three and a quarter years from clay model to first production. These were the first completely new passenger cars the company had offered since the K-car of the early 1980s. The platform team approach meant all team members were located within a five-minute walk of each other. Questions could be answered quickly in a face-to-face interaction. Robert A. Lutz, President of Chrysler, said of the company's achievement:

> What we're finding is that having fewer people and limited resources doesn't have to be a liability – not if those resources are properly focused and not if your people are organized so that they're genuinely turned-on and feel that their contributions really do make a difference. With our platform team setup, we were able to bring this entire family of cars, with all their sophisticated technology, to market in a world-class 39 months. We also did it with a team consisting of just 744 people. That's easily half the number we would have used in the past. From now through the middle of the decade, we're going to be introducing another all-new product at Chrysler every six months – which is something we never could have accomplished under our old system. We believe that without the right kind of culture in place in the future, the management of technology probably isn't going to be very successful.[70]

By 1994, new product development cycles were half of what they had been at the start of Chrysler's reorganization. The JA project, relaunching the company's presence in the compact segment with the Chrysler Cirrus and Dodge Status, went one step further than had the LH project. A product excellence team of eight designers and engineers in their thirties – the target market for the new cars – was created to understand the customer. The team held customer clinics and spent time at shopping malls watching people use their cars.

The company has won accolades from a variety of impressive sources for its new products. The 1993 Chrysler Concorde, Dodge Intrepid and Eagle Vision were selected by the editors of *Automobile Magazine*, a publication which has a global rather than North American orientation, as the '1993 Automobiles of the Year'. In 1994, Tomoyuki Sugiyama, Executive Chief Engineer of Honda R&D Co. Ltd said 'he was impressed with Chrysler's lean engineering and noted that the new Dodge/Plymouth

Neon had fewer door-handle parts than the Accord'. 'They are applying great effort to reduce the parts quantity'.[71]

Despite the successes the platform team approach has brought, the company still needs to improve quality. In the May 1994 J.D. Power survey on initial car quality, not one Chrysler car ranked in the top ten, and the rankings of some models had actually dropped. The company is therefore investing substantial sums in training and is changing the kind of people it recruits as new employees: hirees are now tested for basic skills as well as the aptitude to work in teams.

Chrysler's Relationship with Suppliers

In conjunction with Chrysler's move to platform teams, the company began to establish long-term relationships with its suppliers and involve them in the development of new parts and components. Rockwell International, an expert in suspension components and systems, was closely involved in the development of the Viper and shipped complete, ready-to-install assemblies directly to the Chrysler plant in Detroit. In 1992, J. Douglas Lamb, President of Rockwell International Suspension Systems Company (a joint venture between Rockwell International and Mitsubishi Steel Manufacturing Company of Japan), was quoted as saying:

> Our excellent partnership with Chrysler is a cornerstone of our involvement with the Viper. Chrysler's strategy relies on capable and dependable suppliers and we believe we meet that description. In addition, because Chrysler brought Viper to fruition by using a small, nimble product team, the team called on those suppliers who could make contributions in engineering and development. [W]e furnished engineering, key design services, laboratory work and testing, and prototypes.[72]

The Practice of ESI Spreads Beyond the Automotive Sector

Suffering from large market share losses to more competitive Japanese firms, Xerox Corporation became, in 1985, one of the first firms outside the automotive industry to adopt ESI.[73] Xerox has been widely admired for the comprehensive way in which the company embraced and continues to practise ESI. Since 1985, other non-automotive firms in the US, Japan and Europe have chosen to adopt the practice. These ESI-adopting firms are to be found primarily in assembly-based industries which manufacture relatively complex products such as consumer appliances, consumer and industrial electronics, and office equipment. Here two examples, Motorola and Fuji Xerox, from two dissimilar industries (two-way radios and photocopiers), offer a sample of the different ways ESI has been adopted outside the automotive sector.[74]

Motorola's New Product Development Process[75]

Changes in Motorola's NPD process began in 1987 when the project team working on the Bandit pager project rejected the traditional process in favour of a new-style 'contract book' agreement which gave the team great latitude. The contract book was a series of documented agreements between the Bandit project team and the company's senior management which set technical and financial goals for the project. Both sides committed themselves to these targets.

The aim of the project team was to design the pager for manufacturing and assembly ease. The project team moved its workspace to the production floor, but maintained close links with senior management through a sponsor from the upper ranks. Hewlett-Packard was the main supplier and, as such, was made a member of the development team. The pager was launched 18 months later on the target date agreed in the contract book. Development time had been cut in half compared with previous models. Cost targets were met, and manufacturing quality was very high. The product proved both a technical and market success, and signalled the start of a new era in product development.

Motorola's target for 1992 (set in 1987) was to achieve defect rates of 3.4 per million components manufactured. Although the company missed the overall target, it managed an average of around 40 defects per million which compared very favourably with a rate of 6000 per million in 1991. The goal set for 2001 is to meet a defect rate of one defect per billion by cutting manufacturing defects by 90 per cent every two years throughout the 1990s.

Motorola's Relationship with Suppliers

Motorola began to implement ESI after hearing that Japanese companies made more use of their suppliers. As outlined in Chapter 2's example, the company realized that it was giving suppliers designs that were not readily manufacturable, leading to long cycle times and higher-than-necessary costs. Not understanding supplier processes meant a waste of company time and materials. The company's suppliers were telling them, 'Listen to us before you hand us a design if you want to get a lower-cost design.'

The ESI programme gained credibility through upper management support. The company produced a booklet of ESI guidelines in conjunction with its suppliers. Figures 3.1 and 3.2 are excerpts from this booklet. Each of the General Managers sent out directives to their product engineering groups to the effect that they would adhere to the ESI guidelines. The programme started to take off after a couple of project successes. The engineers saw that they did not have to spend such a lot of time on administration, supplier selection, and rework. The company then started to do little things like giving out joint awards to the Motorola engineer and the supplier. Purchasing moved away from tracking the design (as changes were made) to supporting standard running products and looking after the day-to-day activities of the purchasing function (current quality, current pricing, current shortages, and so on).

Motorola's Responsibility

- Match supplier with part (identify close working relations with sub-suppliers).
- Develop contracts negotiated up-front on joint technology, developments and commitments.
- Provide written documentary assembly information, part function, application, critical design parameters and environment.
- Provide non-disclosure agreement.
- Provide quality, cost, delivery requirements.
- Communicate key development schedule dates for prototypes, pilot, and production material.

Supplier's Responsibility

- Provide risk assessment of selected technologies versus quality, cost, delivery.
- Provide manufacturing input.
- Provide recommendations on material selection.
- Assist engineering in making quality, cost and delivery trade-offs.
- Provide itemized cost estimates for direct material cost, tooling, and development based upon Motorola's inputs.

Fig. 3.1 Division of ESI responsibilities

There are two certain Early Supplier Involvement 'killers'. The first is to award the business to a low bidder after the ESI supplier has made a significant contribution. The second is to rapidly move the business to the Far East or to a low bid supplier.

To stimulate and grow the ESI concept and acceptance with our suppliers is the name of the game. With this thought in mind, these guidelines have been adopted.

First, we need some degree of consistency across the Sector in how we reward ESI contribution from the supplier.

We want the supplier to recover his engineering cost with the business placed. Alternatively, and less desirable, we can compensate him directly for his efforts.

Please recognize competitive pricing is still an issue. However, this does not mean that the contribution from the ESI supplier is to be overlooked.

We need to provide full credit to the supplier for their cost reduction, quality improvement contribution. This may be cost avoidance up-front.

There are a number of alternatives and perturbations in business placement. Here are a few:

Life of part award – this may be combined with an agreement on experience cost and pricing with some agreed annual cost reduction amount.

First year business award – 100 per cent.

An agreed-upon percentage of total business if a second source is contemplated – Far East or a lower cost supplier.

Involve your Division Sourcing or Procurement Manager up-front for his or her assistance and inputs. They can help us on the consistency issue and also bring their experience and judgment to the issues.

In conclusion, we need to address the expected future component sourcing up-front – particularly where ESI is appropriate. With this in mind, we need to develop a plan to appropriately and fairly recognize supplier contribution.

Fig. 3.2 Motorola guidelines for fair supplier treatment for ESI contribution

In order to make ESI work, the company created the Commodity Specialist group which consisted of people with engineering, design and purchasing backgrounds.

The Commodity Specialist group reported to Manufacturing Strategy and looked at all long-term issues that were not getting done. This group aimed to link together the purchasing, engineering and supplier organizations. When an engineer came up with an idea, the group tried to match that idea to the right supplier. Doing so saved the engineer a lot of non-value-added work interviewing suppliers to figure out which one was the best. The Commodity Specialist group pooled knowledge from all parts of the organization to understand the true capabilities of suppliers, and looked after the long-term and supplier recommendations (assessment of capabilities, supplier factory capacities, and so on). The Commodity Specialists now managed the tracking of the design. As these specialists were drawn from Purchasing, they were trusted with the authority to place the initial purchase orders. The company felt that, in the long term, the group might not be needed, but that it was a way to get started on effectively managing the ESI process.

Even by the mid-1990s, seven years after the first ESI project, not all ESI projects were successful. The company still did not bring in suppliers early enough. Ideally they would have liked to have them in at the concept phase, as Motorola sometimes was still too rigid regarding design. The company also did not always take up the suppliers' suggestions. A supplier could still end up giving in to something on the Motorola side, even though the supplier knew that he did not have the capability, and had said so. Suppliers sometimes still overestimated their capabilities (for example, in plastic, the ability to reduce the thickness of pager walls). Some suppliers also remained afraid to tell Motorola that the manufacturer was wrong.

Motorola created a Supplier Advisory Council to manage relations and sound out ideas. The Council consisted of 15 rotating – every two to three years – capable and open suppliers and met with Motorola three times a year. An example of the Board at work was the development of the ESI Project Nondisclosure Agreement. When Motorola initially drew up this agreement the Supplier Advisory Board said it was too complicated. The company said it would re-do it. It gave the revised version to the Board the night before the next meeting. At the meeting the next day, the spokesman for the Council said, 'We're not partners with this,' tore the 15-page document in 100 pieces and threw it in the air. A joint manufacturer–supplier effort followed. The result was a very concise document (Figure 3.3). Without the input of the Supplier Advisory Board, Motorola might have sent out the 15-page document to 250 suppliers, and still not have understood the supplier company's side of it.

Motorola aimed to limit the work the company placed with a supplier to no more than 30 per cent of a supplier's capacity. Unlike Japanese manufacturers, the company did not want suppliers to be too dependent on it. The company tried to find suppliers that could manage prototyping, short runs and mass production runs. Traditionally these had been

This Agreement is effective as of the date of the last signature hereto between the parties listed below.

THE PARTIES

Motorola, Inc.	
....	....
....	....
....	....
"Motorola"	"Supplier"

WHEREAS, Motorola and Supplier have previously executed an Early Supplier Involvement ("ESI") Master Agreement effective _______ setting forth certain general terms and conditions for the mutual advantage of each as regards a Motorola product to be designed; and

WHEREAS, the ESI Master Agreement anticipates the creation and execution of subsequent ESI Agreements for the mutual advantage of each as regards a Motorola product to be designed; and

WHEREAS, this Agreement is intended by the parties to be such an ESI Project Agreement as regards Motorola's future product currently identified for convenience as ______________________________ ("Project")

NOW THEREFORE, in consideration of the representations and obligations set forth both above and below, the parties hereby agree to the following terms.

The parties hereby incorporate by reference thereto, in its entirety, the ESI Master Agreement, previously executed and effective ________ to govern this ESI Project Agreement.

The Confidential Period shall be ________ (__) years from the date of disclosure of the Confidential Information. This provision shall survive both the ESI Master Agreement and this ESI Project Agreement.

This Agreement shall involve certain information regarding the Project, including design approach, technology, and market sales forecast related thereto.

IN WITNESS HEREOF, the parties have caused this Agreement to be executed by their duly authorized representatives.

Motorola, Inc.	Supplier
By: ____________________	By: ____________________
Title ____________________	Title: ____________________
Date: ____________________	Date: ____________________

Fig. 3.3 Early Supplier Involvement Agreement (*source:* Motorola)

separated. Finding one supplier that could handle all three cut down on learning cycles.

As of the mid-1990s, Motorola felt the company was not even halfway there in terms of ESI, but believed strongly in the concept. Management understood that new computer companies, for example, could introduce the ESI culture right away, but in an established company like Motorola, the culture was difficult to change. The internal hurdles were the most difficult to overcome. Engineers – all 4000 of them – had been trained to solve problems. Now the focus was on getting others to help solve and to prevent problems. Suppliers bought into this concept more easily, because it meant business, often long-term business, to them.

Fuji Xerox (Ebina Plant) [76]

Fuji Xerox was established in 1962 as a joint venture between Fuji Photo Film Corporation and Rank Xerox, itself a joint venture between Xerox Corporation, the American company which (under the name Haloid) had introduced electrophotography in 1948, and the Rank Organisation plc of the United Kingdom. The initial mission of Fuji Xerox was to produce Xerox Corporation products under licence and distribute them through rental arrangements. Over the years, the company added other activities to become a manufacturer of photocopiers in its own right by 1973. In 1980, the Deming Prize was awarded to Fuji Xerox in recognition of its excellence in quality control.

By the mid-1990s, Fuji Xerox had become a global player. Japanese sales still accounted for 93 per cent of its turnover although the company had established a number of sales and manufacturing sites throughout Asia. The company had also entered into a joint venture with Xerox Corporation called Xerox International Partners in order to sell low-end printers in North America and Europe, along with a second joint venture manufacturing subsidiary that assembled low-end printers in mainland China.

By 1995, Fuji Xerox was one of the leading players in the Japanese market. Its market share was 24 per cent of units sold behind leaders Canon and Ricoh and 47 per cent in terms of copy volume. The gap between the two market share indicators was due to the fact that Fuji Xerox was not present in small printers but held almost 60 per cent of the high-speed and ultra-high-speed segment.

Various factors determined the evolution of certain product and technological characteristics of photocopiers. First of all, there was continuous pressure to merge several document handling processes: copying, printing and telefax. As a result, copying machines needed to have the appropriate software capability for such functions as storing and combining original documents.

In 1982, the Ebina plant became the home of the newly established 'Corporate Research Laboratories', marking a strong commitment by Fuji Xerox to research and technology. A large number of innovative technological introductions ensued in a variety of areas – from copying and printing, to computing, networking and programming. These efforts demonstrated FX's intent to diversify its activities, while also establishing the reputation of the Ebina plant as the centre of FX's technological excellence.

Ebina's New Product Development Process

Development of new products at Ebina was essentially a cross-functional process which followed a series of 'phases and gates'. While the process appeared to be a sequential one, when a difficulty occurred during any given phase, the nature of the problem would be evaluated and a decision made on whether to proceed at once to the next phase or to wait until the problem was solved. Thus, in practice there was overlap between the different phases. Developing a typical new Fuji Xerox product, that is, one with no significant technology changes, would take around 30 months to complete. But products that required designing a radically new technology might take up to ten years (five years for concept configuration and another five for development).

The 'Programme Team' was another important organizational element in new product development. Typically, the team included representatives from the Development, Engineering, Production, and Sales and Service Departments. The team, led by a member of the 'Product Planning and Programme Management' Department, was responsible for managing the different phases. The Programme Leader was selected from the 'Product Planning and Programme Management' group on the basis of experience: the more complex/strategic the development, the more senior the Programme Leader. However, unlike in the automobile industry, the Programme Leader was not the exclusive 'boss'. Often, there were representatives from Development or Engineering who were even more senior, and decisions were taken by consensus. Programme team members typically had two bosses, with the functional managers generally carrying more weight. The team members only represented their particular functions for the purposes of the Programme; however, decisions made by these representatives on the Programme team were normally accepted by the function heads.

Ebina's Relationship with Suppliers

Suppliers were not overly involved in product development at FX's Ebina plant. By and large, they were contacted only after the product design had been completed (that is, after the prototype phase) to discuss manufacturing possibilities. By 1989, it was felt that this system created unnecessary delays in the development process, especially between the design phase and production ramp-up. Involving the suppliers, particularly in the development of tools early on, was expected to help speed up the process. Xerox, the American parent company, had been implementing ESI since 1985 and by 1988 was actively promoting this concept among its sister companies throughout the Xerox group. Ebina's Engineering Department enthusiastically decided to adopt this approach.

The ESI policy that existed in 1995 at the Fuji Xerox Ebina plant had gradually evolved since the first experiments made in 1989. Three main periods could be distinguished in this evolution process. The first period extended from 1989 to 1991. At that time, Fuji Xerox implemented only one type of ESI activity, known as 'Type A', which was the lowest level of 'Supplier Involvement'. Suppliers started to work with Fuji Xerox in the middle of the 'Feasibility Model' phase, after the drawings were nearly completed. They did not participate in the basic design process at all, and only gave input on the detailed design, usually through the VA/VE activities. At most, they might contribute to drawings of the details, but Fuji Xerox maintained responsibility for the design. The advantage of this approach was being able to apply the vendor's experience for making the design easier to manufacture, thus reducing the need for design changes at later stages in the development process.

This approach was applied extensively, with some 205 parts and 41 suppliers being involved during the 1989–91 period. As a result, the company benefited substantially, particularly in terms of cost reduction. Fuji Xerox estimated that costs were 13.1 per cent lower than in the pre-ESI period. However, there was some dissatisfaction. The initiative fell short of the cost reduction target of 21.5 per cent that had been set. In addition, there were problems of cost increases during the initial production phase due to delays in the implementation of quality measurement. Also, high-assy[77] parts that had been excluded from ESI did not reach their cost target. These problems occurred basically because 'Type A ESI' involved only a single supplier working on design activities and manufacturing the part-subassembly. Therefore, Fuji Xerox decided to revise its approach by promoting competition among its suppliers at the earliest stage of product development.

The second ESI period spanned the years 1992 to 1995. The new goal was to extend the scope of ESI to high-assy suppliers and to establish a 'Cost Management System' from trial to mass production. This objective required the introduction of a new ESI approach – called 'Type B ESI' – to complement (not replace) the previous one. The plan was that suppliers would become involved during the middle of the Feasibility Model phase. The main difference was that two suppliers rather than one became involved (if the required volume was sufficient). They would work jointly through the Feasibility Model and Engineering Model phases, and then they would compete on costs in order to obtain the long-term supply contract. Fuji Xerox did not believe that suppliers might reject this idea although it would require an investment of several man-months' worth of engineering work without being assured of ultimately receiving the order. Fuji Xerox knew that these suppliers did a lot of business with the company (at least Yen 500 million per year). So, if one order did not materialize, they might succeed on another and therefore feel only a limited impact on turnover.

During this second period, nine high-assy units and nine suppliers were involved, and the concept of 'Cost Management Activity' was introduced. Fuji Xerox also initiated a 'Defect Rate Improvement' system. The targets were not only achieved, they even exceeded expectations. Fuji Xerox was pleased with these results, but management felt that there was still some room for improvement. First, it seemed that the ESI approach did not really optimize the suppliers' technological expertise, since they were involved only during the Feasibility and Prototype Model phases, which did not give them enough time to make a significant contribution. Second, Fuji Xerox thought that having the 'competitive' bidding process take place at the end of the engineering phase was not entirely satisfactory, as the winner then did not maintain the same incentive to perform once the competition was over. In practice, it was important that costs continued to be reduced at the mass production level. Third, Fuji Xerox noted that ESI needed to be extended to the parts procured overseas, which represented a growing proportion of the process.

The third period is to cover the years 1996 to 2000. Essentially, the objective is to advance the ESI starting point to the concept/product planning phase in order to achieve better cost targets. The selected suppliers would thus contribute before the development of the 'Bench Model'. Building on the experience of the second period, which had proven so successful in terms of cost reduction, Fuji Xerox wants to pursue the concept of having competition between two suppliers for each part/high-assy unit. Thus, Fuji Xerox will continue to invite two suppliers to participate in the concept/product planning, basic design and production design phases. Then, both suppliers will receive orders for production, thereby actually staying in the competition during the product life cycle. Fuji Xerox calls this approach 'Type C ESI'.

'Type C ESI' is expected to ensure optimum cost reduction, thanks to the manufacturing suggestions by suppliers that are incorporated in the part design as well as through continuous price competition between suppliers throughout the product's life. The two suppliers involved in 'Type C ESI' are always under pressure to reduce costs. In addition, management believes that involving suppliers early on would help them to reach higher quality targets for manufactured parts. This new approach to ESI is not meant to replace the previous ones. Rather, it will complement them and be applied to parts or high-assy units whenever it is appropriate (see Figures 3.4 and 3.5).

Conclusion

The examples described in this chapter not only outline the origins of ESI, but also the diffusion and evolutionary paths the practice has taken in different organizations and regions. Whether these differences are most strongly affected by industry, geographic location, management choice or economic conditions is the subject of Chapter 4. It does seem that the dynamics play out differently depending on the mix of these factors.

What is clear at this point is that ESI's origins as a systematic practice date back to the Japanese automotive industry of the late 1940s (see Figure 3.6). By the late 1970s, the practice was widespread in the Japanese automotive sector, and by the early 1980s had moved to the US with the Japanese automotive transplants that were just beginning to be established. An increasing range of American manufacturers, including Xerox, Chrysler and Motorola, were working to implement the practice by the late 1980s. Shortly thereafter, European manufacturers, both automotive and non-automotive, began to experiment with ESI. The practice had moved full-circle back to Japan by the mid-1990s, with some very sophisticated versions being implemented by some of that country's non-automotive manufacturers.

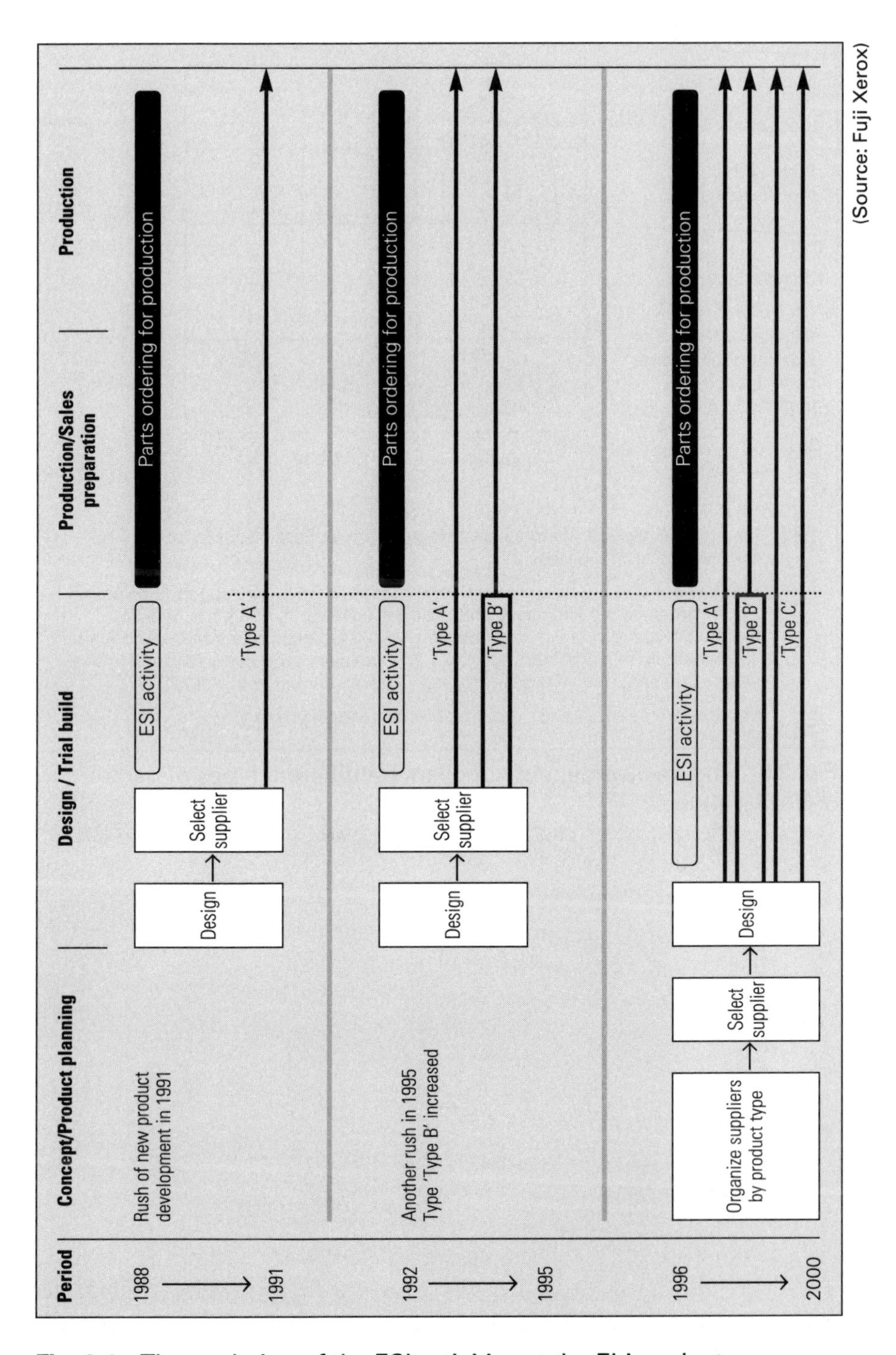

Fig. 3.4 The evolution of the ESI activities at the Ebina plant

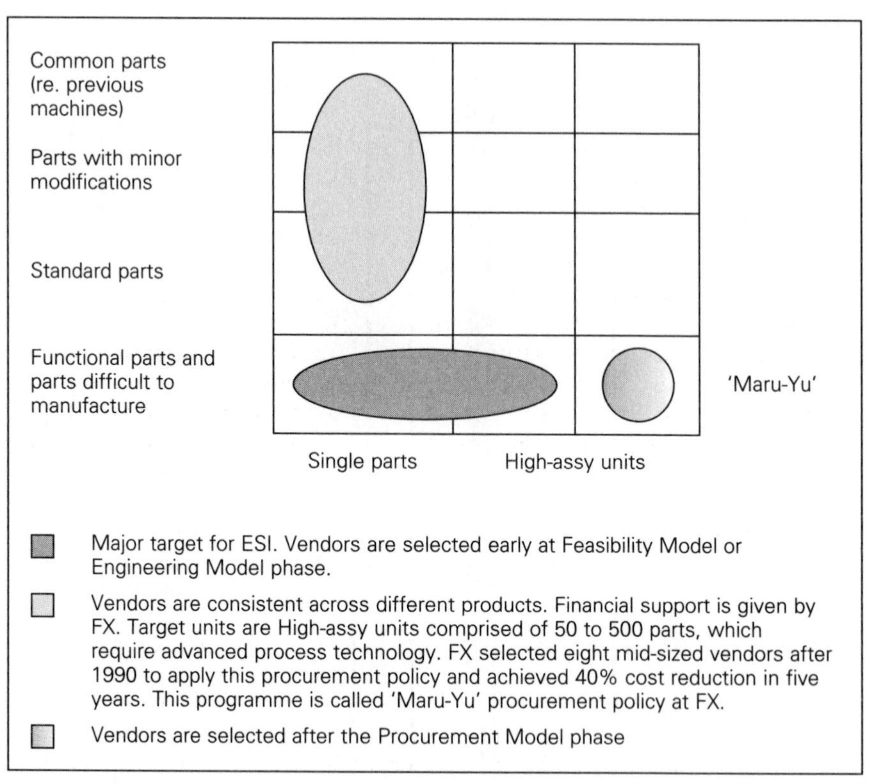

Fig. 3.5 The involvement of suppliers for different types of parts
Source: Fuji Xerox

Pre-Second Word War: Some suppliers to the locomotive and aircraft industries are probably involved in the development of 'black box' parts. The process does not appear to be systematic and/or widespread. Some of these suppliers, such as those manufacturing gears, start to work with the new automotive manufacturers.

1949: Nippondenso is spun off from Toyota as the auto firm's quasi-independent first-tier supplier of electric components. Toyota's electrical engineers become employees of Nippondenso. These engineers apply the same 'black box'-style approach to part design as a supplier to Toyota as they had been used to doing as employees of Toyota.

1950s to 1970s: Toyota develops 'lean production' which includes the practice of ESI. Strong supplier groupings, or *kyoryokukai*, are formed. Lean production takes a full 20 years to be implemented throughout the company's supply chain. Other Japanese automotive assemblers, including Honda, copy Toyota's system. However, Honda does not create supplier groups.

Late 1970s: Honda Japan starts to develop and implement its B/P system with its Japanese suppliers.

1980s: Japanese automotive suppliers start to make engineering proposals and conduct VA/VE.

Early 1980s: The first Japanese auto 'transplants' are established in North America.

1982: Honda US begins car production at Marysville, and gradually starts to implement the company's B/P system with its new American suppliers. The process is not smooth, nor complete more than a decade later.

1985: Xerox Corporation implements its version of ESI which the company calls CSI (Continuous Supplier Involvement). It is possibly the first non-automotive company to institute the practice.

Late 1980s: Chrysler starts to adopt 'Honda-style' new product development methods, including ESI. It is the first indigenous American auto assembler to do so.

Late 1980s/Early 1990s: An increasing range of non-automotive manufacturers start to experiment with ESI. The practice is seen in industries such as consumer appliances, consumer and industrial electronics, and office equipment, as well as in North America, Europe and Japan.

1990: The Honda Accord becomes the first best-selling non-GM or Ford American-built car in the history of the country's auto industry.

1993: The Chrysler Concorde, Dodge Intrepid and Eagle Vision are selected by the editors of *Automobile Magazine* as the '1993 Automobiles of the Year'.

1994: Lexmark and Minco win the single part category of the Structural Plastics Conference for their co-development of the Liberty frame.

1996: Fuji Xerox starts to implement 'Type C ESI': the concept of competition between two ESI suppliers for each part/high-assy unit throughout the product life cycle.

Fig. 3.6 The story of ESI

4 The Drivers of ESI Adoption

Lexmark: from Selectric to Laser[78]

Lexington, Kentucky is the original site of IBM's typewriter division and home of the world-famous IBM Selectric. Between 1970 and 1980, in the face of relentless foreign competition, the Lexington site went from being IBM's most profitable division to its least. On the assumption that corporate overhead allocations were at least partly responsible, Lexington's management asked IBM to turn the group into a subsidiary and give it a free hand in running the business. This coincided with a period of 'non-core' divestiture at IBM so that, with Lexington's agreement, IBM went a step further: it sold the operation. Lexmark thus became a legal entity in March 1991. The name Lexmark was derived from 'lexicon' (pertaining to words) and 'mark' from the company's saying that it was in the business of putting marks on paper.

Personal laser printers comprised an important product group for Lexmark. In 1991, the US personal printer market was worth $7 billion and forecasted to grow at a rate of 9 per cent per year in volume terms until 1995. Competing directly against Lexmark's entry-level products were a host of laser, ink jet and impact printers from domestic and foreign competitors. One of its chief competitors was Hewlett-Packard, which had revolutionized desktop printing in 1984 with the world's first desktop laser, itself the result of a 1975 alliance with Canon of Japan. HP had quickly vaulted to a leadership position and its position remained unchallenged through the 1980s.

IBM had entered the desktop laser printer market in late 1989 with its 4019 laser printer. Made entirely within the company, the 4019 listed at $2595 and was the only desktop laser produced entirely in the United States. It had half as many parts as HP's Laserjet II launched shortly before at $2695 as well as faster speed, more standard fonts and a comparable 300 dpi resolution. The 4019 was an excellent first product and a potentially significant challenger to the market-leading Laserjet II.

Yet almost immediately, HP introduced the Laserjet IIP which carried a list price of $1495 but sold in many retail outlets for around $1000. Since IBM had no comparable product under development, it countered by de-tuning the 4019 (cutting its speed from 10 to 5 pages per minute) and matching the Laserjet IIP's price. The 4019 5E was introduced in March 1990 and was an obviously poor solution: the need was for a truly new product that would retail at $1000. IBM launched a development effort but by December 1990, its design team announced that the goal was impossible given expected costs.

The stakes were high. If it failed to make a bid in this product category IBM would have rough going in an increasingly important segment of the printer market. It was under these conditions in late 1991 that Greg Survant, Lexmark's programme manager for the Entry Page Printer Group, received a mandate to 'achieve the impossible'. IBM's time-honoured product development process – broadly similar to that of many other American manufacturers during the 1980s, as noted in Chapter 2 – seemed an obstacle to Survant:

- First, a design group – typically engineers – conceptualized a new product based on technological capabilities.
- If shown to be profitable on a materials cost basis, the product would move to a final drawing stage and then be 'thrown over the wall' to manufacturing.
- Manufacturing then determined how to produce the product efficiently and effectively and only unavoidable changes were sent back to the design team.
- Parallel to manufacturing's work, purchasing tendered for subcontracted parts.
- The selected suppliers then manufactured parts and subassemblies according to the specifications supplied, and (hopefully) delivered on schedule.
- Various tests were carried out during the development process and once successfully completed, mass production began.

By its very nature, this was a lengthy process that tended to produce rework and trouble-prone production. The 4019, for example, had been plagued by emergency and ordinary engineering changes and had taken 36 months to reach mass production. Survant believed that his 'impossible' mandate – to develop in 18 months the 4037 5E or 'Liberty' printer, an under-$1000 desktop laser that would compete with HP's new offering – required a different approach. One feature of this new approach involved supplier relationships which, under IBM, had been adversarial: potential suppliers were given final specifications, submitted their bids, and the lowest price usually won the order. Together with other innovative ways of moving the project forward, Survant came to view the issue of supplier relationships as critical to the Liberty's and Lexmark's success.

The previous chapters discussed the benefits that can be obtained from an effective ESI relationship: companies report that their development times are shorter, their costs lower and overall product quality improved. Indeed, these and related benefits have been amply documented by car manufacturers where the expectations for ESI have grown in relation to the industry's experience with it. But ESI is not a universal phenomenon: some companies adopt it, others do not. If ESI represents best practice for some of the world's most successful companies, shouldn't they all be developing ESI relationships with major suppliers?

Lexmark's development dilemma with the 4037 is a case in point. IBM's promotion of the American-classic PPP (project planning programme) model of sequential product development had achieved impressive results during the 1970s and 1980s, and while foreign competitors were developing ESI relationships IBM never considered doing so. The PPP tradition had deep roots in Lexmark's corporate ideology but the new company also found itself navigating in strange waters: the competition, both domestic and foreign, was developing better products at a faster pace than during IBM's tenure and putting severe downward pressure on retail prices. Lexmark found it necessary to move from PPP to ESI and hindsight argues that had it not done so, the company would have had a difficult time with what quickly became a fast-paced, high-quality, price-conscious and customer-sensitive market for desktop lasers. Is it possible to identify the factors that led to this change of direction? Can we discern certain features of the external environment and its internal operation that led Lexmark toward ESI?

There are, it turns out, a number of factors that lead some firms to adopt ESI and others to ignore it. In this chapter, we develop a model of the factors that lead some firms to adopt ESI. These factors are both external and internal to a firm and we have termed them the 'drivers' of ESI adoption. The chapter begins by asking what characterizes a company that considers ESI, then develops the model of ESI drivers and closes with the implications for practising managers.

The Characteristics of ESI Adopters

Lexmark adopted ESI although IBM had always taken a different path. Philips Floorcare adopted ESI in 1993 while Hoover began to consider it in 1994. Toyota was involved with ESI in the 1950s; GM first learned about it in the late 1980s. What accounts for these differences? Isn't it logical that a close and stable relationship between a manufacturer and a supplier benefits both? If so, why are some hurrying to adopt ESI while others show little interest? The answers to these questions appear to have several dimensions.

First, companies interpret their environment differently along with the information it holds. Were it possible to compare the environmental surveys from five different companies operating in the same industry, there would be interesting and notable divergences. Different companies also operate with widely varied internal resources, factors and constraints. Putting these two together – different interpretations of the environment linked to contrasting sets of internal realities – we obtain a cluster of variations between companies that forms part of the answer. A second important element involves differences between environments themselves. The industrial and market realities of printed circuit manufacturing differ considerably from those of furniture, for example, and European production of either differs from arrangements in Japan or North America. The operating environments of companies vary across many dimensions, some concrete (such as regulatory structure) and others more abstract (such as the level of technological intensity).

If we use the concept of ESI as a lens to magnify this diversity – that is, to focus on firms that have adopted its operating principles – it becomes possible to extract a set of distinguishing characteristics. Over a cross-section of industries and across geographic boundaries, we have found that the following conditions are often associated with companies that adopt ESI:

- **Heavy competition** ESI-adopting companies tend to operate in highly competitive markets. Typical examples include automobiles, consumer electronics and advanced office equipment (for example, copiers).
- **Fast-paced environments** Time is a factor for all companies but it seems particularly important for ESI adopters. Companies report a growing need to shorten their development times in order to stay competitive.
- **Price-sensitivity** Cost is also a factor for all companies but again, ESI adopters typically report scenarios that make lean operating budgets an imperative. These companies tend to be in markets where product pricing faces continuous downward pressure and they find it necessary to continually improve their operations in order to position products competitively.
- **Growing technology mix** Products are becoming increasingly sophisticated for ESI adopters and sophisticated products tend to be complex in two ways: (a) they embody a growing variety of technology

and/or (b) the technologies they employ have become increasingly complicated or specialized.

- **Product proliferation** Companies report that their product lines are growing in size: the number of products offered is expanding along with the number of features offered per product, increasing the number of models put on the market.
- **Growing operational complexity** As products proliferate and become more complex, internal manufacturing systems become larger and more resource-intensive. Products that incorporate complex components and subassemblies are placing heavy demands on engineering and manufacturing resources.
- **Strained internal capacities** Some ESI-adopting companies flatly state that they lack the internal engineering or design resources (time, people, expertise) to accomplish a project. Product proliferation and the spiralling increases in technology mix strain the internal capacities of companies.
- **A focus on core competencies** Companies adopting ESI have generally made a strategic choice to focus their energies and talents on a core set of distinctive competencies. This runs in parallel with the observation that these companies tend to be moving away from vertical integration and toward the assembly of products for which they develop and control the overall design.
- **Competent and reliable suppliers** ESI-adopting companies are in one of two situations: either they have access to a competent supplier base, or they are in a position to develop one. Both scenarios are in evidence: some companies are approached by expert suppliers who propose design-level involvement, and other manufacturers work with suppliers to develop their internal competence. The availability of competent suppliers that can meet production requirements is an obvious necessity.
- **Supportive norms** Companies adopting ESI hold norms that encourage collaborative business practices; these norms are found within national cultures, industries and professions. The Japanese culture, for example, is frequently noted for its emphasis on trust and relationship; both of these are key underpinnings of an ESI relationship and the practice has flourished in that country. Industries have norms as well. In the manufacture of aircraft, for example, it has become accepted practice to involve outside experts.

Taken together, these ten factors characterize companies that adopt ESI: they are in heavily competitive situations, time is at a premium, cost is a key concern, the technologies they rely on are increasing in number and complexity, the size of product lines is growing, operations are becoming more complex, and so on. The scenario may seem familiar to many managers but the ESI-adopting executives we interviewed in our research were nearly unanimous: their experience of these forces was so acute as to demand action. They reached a breakpoint, so to speak, beyond which business-as-usual was simply not viable.

Characterizing the forces that stimulate ESI adoption is helpful, but at this level the decision-making map contains little detail. In order to develop more precise and useful knowledge, we launched a research programme in 1994–95 which sought to identify a set of underlying and generative factors. This research distilled a decade of ESI-related investigations (generally in the automotive setting), synthesized that information and developed a blueprint of the important factors, and then investigated the relevance of those factors in the general business arena. Once assembled, the research results provide us with a model of the factors that 'drive' ESI.

The 'Drivers' of ESI

It has been said that there is nothing so practical as a good model or theory. The first step in developing a good model is categorization: the various phenomena are examined with an eye toward grouping them according to common attributes. Looking at the list in Table 4.1, it is clear that some of these characteristics describe conditions that are external to the organization (competition, technology mix) and that others are clearly internal, relating to strategic or operational issues and choices (operational

Table 4.1 Characteristics of ESI-adopting firms

Heavy competition
Fast-paced environments
Price-sensitivity
Growing technology mix
Product proliferation
Growing operational complexity
Strained internal capacities
A focus on core competencies
Competent and reliable suppliers
Supportive norms

complexity, core competence). This is logical and leads to the first conclusion: both internal and external factors seem to be leading companies toward ESI. This is consistent with current research, but while some investigators have reached the same conclusion there are others who focus entirely on one or the other set of factors. This, we believe, is a mistake, for it seems clear that the internal and external factors in play are almost inextricably enmeshed.

It is possible to make a finer distinction with regard to external factors: there are those that are objective or 'hard', such as the undeniable speed of technological advancement, and others that are more subjective or 'soft', such as the influence of cultural norms. We believe it is useful to differentiate these two and set apart social and cultural factors from more objective ones. In so doing we arrive at a set of three major categories that appear to influence firms toward ESI adoption (Figure 4.1). The external environment exerts a number of influences, as do social and industry norms, as well as the organizational choices a firm makes regarding the nature of its operations.

This general model allows us to take the next step, which involves the identification of specific issues in each category. By articulating these, we believe that executives gain some control over their operations and researchers will be able to develop a more robust body of knowledge. The following section discusses the issues in each category that we have identified through academic and on-site research.

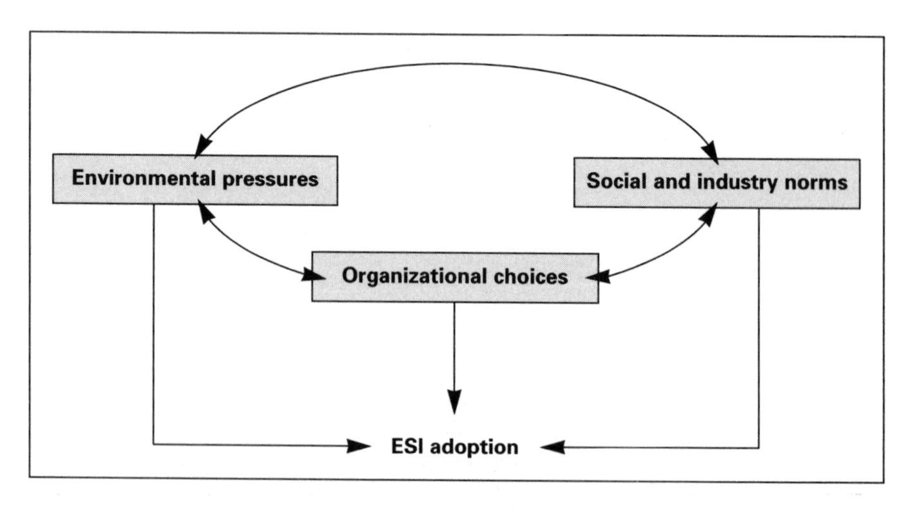

Fig. 4.1 Three categories of factors influencing the adoption of ESI

The External Environment

Broadly defined, the external environment is the general economic, political, cultural and technological context in which all organizations are embedded. Several decades of research on ESI in the car industry has developed a set of environmental factors that are widely viewed as encouraging ESI (Figure 4.2). Some researchers go so far as to state that these pressures are determinate and others only secondary.[79] We will leave this question aside for now and simply discuss the factors that have been well documented.

Competition

Of all the environmental factors noted here, the pressure of competition is perhaps the most global but also the most frequently cited by both managers and academics alike. Competitive pressures on ESI adopters appear to be unrelenting and intense: companies report being forced to continually develop new ways of staying abreast of the competition. Research programmes launched during the 1980s, for example, showed that Western automotive companies developed partnerships with their suppliers in order to deal with the competitive challenges mounted by Japanese car manufacturers.[80] Competitive pressures take many forms, but those most frequently mentioned include the pace of new product

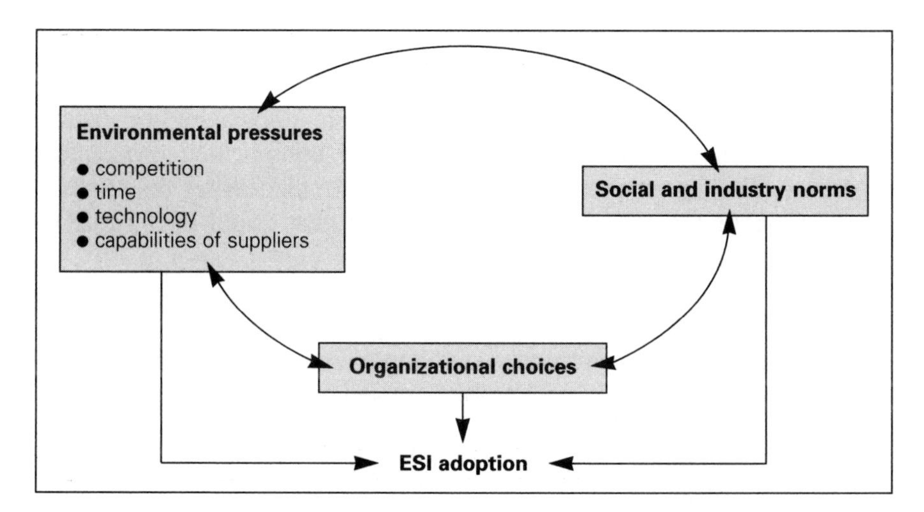

Fig. 4.2 Environmental factors

introductions and product enhancements, the introduction of new innovations that force companies to stay on the cutting edge of their industry, and downward pressures on price.

Time

Most industries are confronted with the need to decrease development times, as faster product development processes and more rapid market introductions provide a crucial buffer against competitive pressures. This is especially true in assembly-based industries. Western automobile manufacturers, for example, reduced their lead-times from five years in the 1980s to under 3.5 in the 1990s. Similarly, the average development time for a vacuum cleaner was three years in the last decade and today the industry is approaching 18 months. ESI allows companies to shorten not just their tooling and development times, but also manufacturing time since an ESI-developed product has fewer parts and more supplied components. The common driver is time. Companies faced with shortening their lead-times on a continuous basis are, in effect, faced with the reality of ongoing process improvement and redesign. The rate of technological change and the need to stay competitive in constantly evolving markets is a serious challenge.

Technology

The pressures of technology exist in several forms but research in the car and electronics industries[81] has specifically identified two: the growing variety of technology and the complexity of production that originates from product proliferation. This research has shown that each new generation of products incorporates more and increasingly complex components that in turn require increasingly complicated production processes. Both generate a heavy demand on engineering resources.[82] There are four technology-related influences that could lead firms to adopt ESI:

1. Products are incorporating an increasing variety of technologies.
2. Each technology is continually evolving to a more sophisticated state.
3. The growing mix of technologies used in a product increases production complexity.
4. The technologies indigenous to an industry may be particularly suited to ESI practices.

Capabilities of suppliers

ESI is only feasible if suppliers possess the competencies a manufacturer requires. These competencies can be obtained in one of two ways: they may be found ready-made among an existing group of suppliers, or they may be purposely developed by the manufacturer. Many large plastic moulders, for example, now possess levels of expertise that outstrip those of their manufacturing clients. US car manufacturers, on the other hand, found it necessary to 'develop' their supplier base in order to promote advanced supply relationships during the 1980s. In both cases a high level of supplier involvement is influenced by the structure of a manufacturer's subcontractor base.[83] American product development specialist Bernard N. Slade underscored this when he observed that advanced technology is forcing companies to make better use of their suppliers' expertise.

These four factors – competitive pressure, compressed time, complex technology and external sources of expertise – are causing many companies to re-evaluate their product development strategies and seek alternatives. When ESI is incorporated as part of their response, it becomes possible to distribute and effectively lessen the pressures of competition, time and technology across a broader set of resources.

Social and Industry Norms

A norm has been defined as '... an idea in the minds of the members of a group, an idea that can be put in the form of a statement specifying what the members ... should do, ought to do, are expected to do, under given circumstances'.[84] The key to understanding the potential impact of a norm lies in understanding the expectations it carries: all norms prescribe 'correct' behaviour. Because they are an important part of the human psyche, norms are also ubiquitous and exist on various dimensions. Some can be located in the cultural strata of a country – those that are taught to children by families and schools – while others can be identified in a specific place of employment and shown to govern acceptable behaviour in work groups (Figure 4.3). In each case, the norms in play communicate expectations and have a powerful effect on the behaviour of individuals and, therefore, a company.

National Culture

It has been suggested that Japanese culture predisposes Japanese employees to work together in a partnership fashion and that this leads to cooperative relationships between manufacturers and suppliers: because

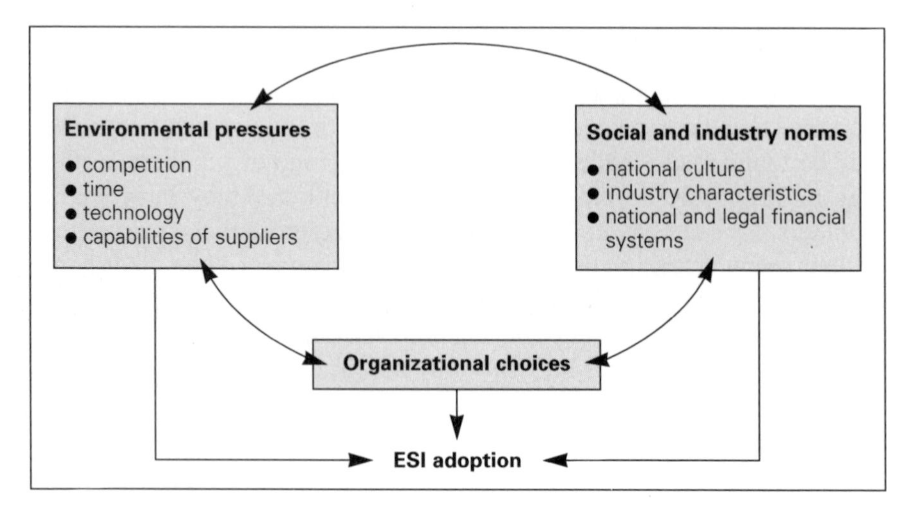

Fig. 4.3 Social and industry norms

of their cultural norms regarding honour, trust, informality, commitment and relationship, ESI seems a logical practice for Japanese companies.[85] Although these norms arise from a socio-cultural base they cannot help but permeate educational and work activities. One Japanese researcher, by contrast, has shown that Japanese manufacturer–supplier relationships were not always so collaborative. They were, in fact, adversarial up to the early 1960s, and evolved to cooperative arrangements only because intense innovation and product proliferation led manufacturers to seek the capacity and expertise that suppliers could provide.[86] The dominant logic of buyer–supplier relationships did indeed become joint problem solving as opposed to adversarial bargaining, but more for industrial reasons than cultural ones. Nonetheless most observers conclude that Japanese cultural norms, centred on collectivism, have a significant influence on ESI practices in that country. Most also agree that the norms of individualism and competitive bargaining which dominate Western market economies tend to work against the practice.

Industry Characteristics

Norms exist for industries as well as national cultures.[87] Some industries, for example, have low levels of contractual formality relative to others for reasons that cannot be explained by economic rationality (for example, the construction industry versus distribution). Others have long traditions of linking multiple expert suppliers together in the assembly of large and complicated products, such as in aircraft manufacture. It is also understood that the norms in an industry can be influenced by its level of international competition and globalization. To the extent that an industry is global, its best practices tend to be emulated across national borders. Western car companies, for example, established partnership arrangements with their suppliers during the mid-1980s as a way of responding to Japanese competition. They did so in order to emulate best practice, which is to say, Japanese practice. But they were only partially successful because, according to one authority, they imported these practices into a cultural context that was radically different from Japan's, and where the national norms were not conducive to the same relationships.[88] Japanese manufacturers and suppliers had a longer history of cooperative relationships, while in the West the free market is founded on norms of competition and adversarial bargaining.

National Legal and Financial Systems

Arising from the pervasive influence of cultural norms, regulatory systems governing legal transactions and financial arrangements are designed to guide the behaviour of individuals and organizations. These systems represent a more formal, codified set of prescribed behaviours. Sako (1992) has documented the way Japanese legal and financial systems facilitated the development of cooperative arrangements between manufacturers and suppliers, whereas Western systems effectively reinforced individualistic and competitive behaviour.

While not as 'hard' or objective as the preceding set of drivers, social and cultural norms do shape attitudes and they can often be detected at the bases of systems and structures that guide behaviour. Incorporating both constraints and inducements, norms encourage one set of actions, deter others and move still others into the realm of unreality. Because norms in an organization's social environment shape the knowledge, beliefs and attitudes of people (and therefore, those of the organization) many researchers believe they have a significant impact on ESI practices. Managers should similarly consider them powerful factors that need to be taken into account when making decisions.

Organizational Choices

The actions taken by organizations often express a healthy measure of discretion and it is due to this that the significant role of strategic choice is generally recognized. This is a departure from the determinist view that regards the practice of management as wholly shaped by manifest external forces which act on companies in irresistible ways. In this view, managers employ a classical strategy-making process to arrive at a set of decisions about the nature of their industry, the nature of their internal operations, and the optimal way forward. In reality, all companies undertake strategic planning through a process that combines rationality, opportunism and serendipity. Certain strategic decisions reflect organizational preferences that are not always preordained by the environment. Four of these preferences have a bearing on the issue of ESI (Figure 4.4).

Level of Integration

The level of vertical integration a firm adopts is a strategic choice and is one of the most frequently noted correlates of ESI adoption.[89] This is entirely logical. To the extent a firm chooses to be vertically integrated, its need for ESI diminishes. But there actually appears to be a trend away from vertical integration, where firms programme all their development activity internally, and one toward vertical cooperation where partnerships are formed within networks of external expertise. In fact, the research on

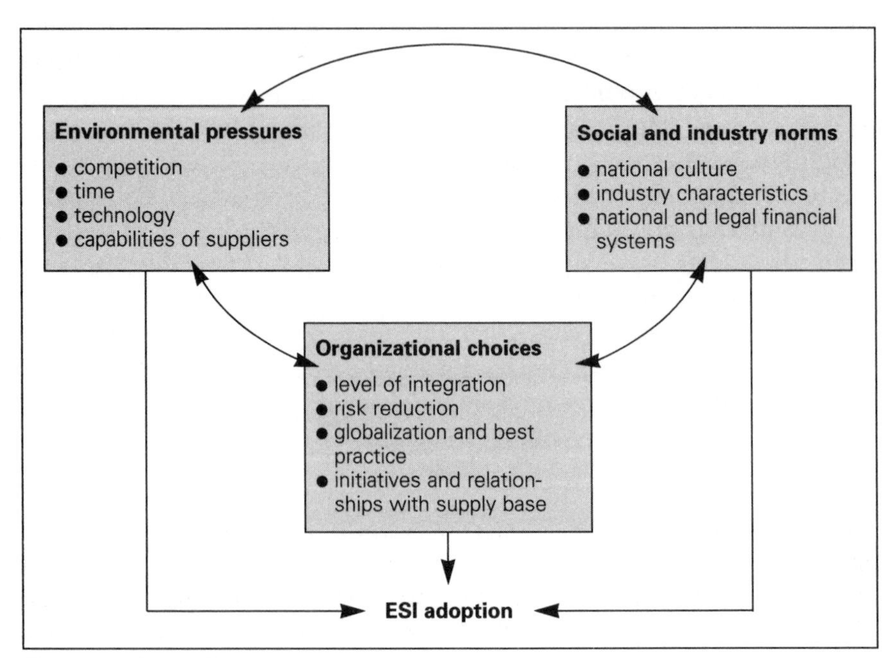

Fig. 4.4 Organizational choices

technology development increasingly describes the innovation process as occurring within a network of actors, among which suppliers are key players.[90] Many companies engaging in new product development are therefore balancing the costs and benefits of maintaining an in-house development capacity with those of relying on external sources. In place of the classic 'make or buy' decision, they can now choose between 'make or cooperate'.

Risk Reduction

Risks defined in terms of resource outlays (personnel, time, financial, physical plant) and eventual product liability concerns are significant factors in many companies' development decisions. The innovation process itself and the development of new products that incorporate complex subassemblies or components place heavy demands on internal resources, thereby raising the manufacturers' risk of poor performance and investments out of proportion to potential returns. Companies may chose to hedge these risks through joint development. One of Kodak's divisions, for example, is under heavy pressure to reduce retail prices and pare down cycle times as much as possible. This division produces Kodak's cameras, which have more than 50 per cent of their value provided by external suppliers; they instituted an ESI programme in 1988 which involves suppliers in the free exchange of information: costs, benefits, risks and returns. Before becoming involved in ESI, one of the division's suppliers of glass lenses charged prices that reflected the potential risk of cancellation and the costs of tooling. In a series of meetings, Kodak and the supplier shared sensitive information which led to much lower prices, and Kodak guaranteed to contract with the supplier for a certain quantity of lenses. Risks on both sides became absorbed in the new dynamic of trust that developed in the relationship.

Globalization and Best Practice

There is an increasing tendency for firms to benchmark the practices of leaders in their own and other industries, and to do so on an industry rather than a national basis. In effect, this formalizes the process of emulation that has existed in business for a very long time. One analysis of the Japanese reliance on the 'black box' system – a transaction where a parts supplier executes the detailed design of a component based on specifications provided by the assembler – concludes that its initial diffusion is better described as emulation than rational calculation.[91] Once begun by Toyota, the practice became widespread as knowledge concerning black box practices was transferred first between manufacturers, and then on to their suppliers.

Initiatives and Relationships with the Supply Base

A company may find itself in a network of highly competent suppliers who are ready to engage in development activities immediately, or it may operate with suppliers that lack the necessary expertise. In both cases a company's priorities toward purchasing and supplier base management are important factors in ESI adoption. When competence already exists in the supply base the company's decision is whether or not to access it, at which stage, and at what depth in the development process. When the competence does not exist among suppliers, the company has the choice of developing it. It may come as a surprise to some Western managers to learn that Japanese manufacturers invest considerable effort preparing their suppliers for ESI. Their suppliers are carefully groomed and selected since only a minority (even among so-called first-tier suppliers) are able to add value in the design process.[92] Developing the supply base in the near-term can often help a company gain the longer-term advantages that ESI has to offer. There are several ways in which this is done, examples of which can be found in the section on Honda's ESI approach in Chapter 3.

In both cases there are choices to be made regarding the type of relationship the manufacturer will have with suppliers. In the classic tendering approach to supply base management, relationships are based on secrecy, competition, price and short-term advantage. The power dynamics of the relationship are kept purposely off-balance, usually in favour of the buyer or manufacturer. This represents a choice even though it may have been made through tradition and default. ESI requires a reverse logic: a relationship founded on trust, cooperation, long-term advantage and mutual benefit. Companies on both sides of the transaction must consciously choose to maintain open and communicative relationships, sometimes to the point of divulging sensitive financial and proprietary design information. ESI requires raising these issues to a more conscious level in order to decide what sort of relationship is appropriate for both firms.

The Critical Factors

Taken together, these three categories of ESI drivers articulate the most important factors identified through research and management practice. We know that successful product innovations make use of outside technology and that either manufacturers or suppliers can assume the lead role.[93] The innovation process is increasingly recognized as occurring within a network of actors, among whom suppliers are increasingly important[94]. Considerable research in the automobile industry has confirmed that the success of Japanese car manufacturers is related to the use of external expertise, an expertise which is lodged within the structure of a supply base and often deliberately developed.[95] A significant trend toward vertical cooperation has been recorded and networks of expertise seem to be on the rise. Apart from general agreement at this level, however, opinions diverge as to which factors lie at the heart of ESI adoption and which remain peripheral. Arguments have been ventured that support the amorphous idea of culture as fundamental, since its various manifestations diffuse norms that predispose people and companies to engage (or not engage, as the case may be) in cooperative relationships. Some analysts of industry structure disagree with this perspective and point out that while overall transaction costs may be lowered through partnerships with external agents, innovation processes are inherently risky and impose heavy demands on internal resources, and that therefore some industries are economically predisposed (in the current environment) toward low levels of vertical integration. Others observe the more pedestrian outcomes of emulation and best practice, saying that the intensity of competition leads companies to benchmark industry leaders in an effort to stay abreast. Still others contend that the proliferation of products, the fact that these products are becoming more sophisticated, and that they incorporate an increasing number of specialized technologies, all lead manufacturers to seek outside sources of competence.

Throughout 1994 and 1995 we conducted a research programme that explored these various propositions. Our goal was an initial test of the most common claims in each category to determine which have a significant effect on ESI adoption. Because there have been few studies of ESI outside the automobile industry, and to determine if and how ESI is being adopted elsewhere, we focused the research on three assembly-based industries: electric appliances, consumer electronics and office equipment. The programme was carried out in three areas of the industrialized world: Europe, the US and Japan.

The Research Process

First, we conducted interviews within 25 manufacturing companies to understand the nature and extent of supplier involvement in their product development process. Our meetings included the project leader of a new product development programme as well as the purchasing manager who was a member of the project leader's team; this provided two key perspectives on the buyer–supplier relationship in question. The focus of discussion was a product development project, not the company as a whole. These interviews were held with seven manufacturers in Europe, nine in the United States and six in Japan. We also met with one car manufacturer in each region to obtain a sense of the differences between our interviewees and their counterparts in the automobile industry.

From these interviews we developed a questionnaire and sent it to each of the executives who had met with us, asking that they again focus on a product development project and characterize the business context, the type of supplier contribution, the management of relationships, and the way this particular buyer–supplier relationship compared with normal process in the business unit. We organized and then operationalized the research propositions noted above in the following way:

Environmental Pressures

- The intensity of competition takes many forms but one is especially noted in the literature: the ever-increasing pace of product innovation. Our measure was the reduction in product development lead time.
- The pressures technology exerts also exist in many forms but again, research tends to underline two: the growing variety of technology and the complexity of production that originates from mass production and product proliferation.
- We measured the changes occurring in technology mix and the number of units typically purchased, presumed to be a measure of manufacturing complexity.
- We also assumed that the extent of liability associated with parts or components purchased influenced (negatively) a manufacturer's choice to implement ESI.
- Finally, the technological base and manufacturing processes indigenous to an industry are also important considerations since certain factors, such as asset specificity, can vary widely across industries. We assessed the industry environments of the respondents.

Social and Industry Norms

- Companies may move toward ESI because of a behavioural norm in the cultural environment for openness and cooperation. We assessed the geographic origin of these companies.

- Behavioural norms also exist in industries. We assessed industry.
- Norms in an industry may be influenced by its level of international competition and the emulation of best practices. We assessed the scope of competition.

Organizational Choices

- The literature specifically details strategic choices that firms make affecting ESI: the level of vertical integration and relationships with the supplier base. The first factor was identified through the perceived level of integration relative to competitors and the proportion of parts used to those that are purchased (purchasing ratio).
- The second was measured by the number of initiatives taken to improve one's supplier base (accreditation, consolidation, JIT programmes and so on) as well as the scope of activity in the supplier base.

Next, we developed an ESI index which accounted for the fact that a continuum of practices fall within its purview, as defined in Chapter 2. We measured these various forms of involvement as follows:

Level 1 The supplier provides input into a product's design by sharing information on equipment and capabilities.

Level 2 The supplier provides feedback on design including suggestions for cost and quality improvements.

Level 3 The supplier participates significantly in the design of a part or component by executing detailed drawings based on a client's rough sketches.

Level 4 The supplier takes full responsibility from concept to manufacture for the design of an entire part or subassembly.

Level 5 The supplier takes full responsibility from concept to manufacture for the design of a system or subassembly incorporating one or more parts which the supplier also designed.

The ESI index was comprised of these five levels as well as the frequency and formality of supplier involvement. Table 4.2 shows the distribution of responses for each item.

For research purposes, then, Figure 4.5 represents the model of ESI adoption that was tested in our research programme. Note that the interaction between categories was not assessed.

The survey was constructed so that the influence of each factor or variable in the model of ESI adoption could be tested to determine its effect on the companies participating in this research. This was done by analysing the extent to which ESI Index scores were affected by a given factor, and observing whether the resulting effect was the one expected. For example, if an executive responded that the level of vertical integration was low for his product development project, we expected a relatively high

Table 4.2 Variables measuring ESI involvement

	Frequency (%)
Levels of ESI involvement	
Level 1: Supplier provided input into your product's design by sharing information about its equipment and capabilities	4
Level 2: Supplier provided feedback on your design including suggestions for cost and quality improvements	54
Level 3: Supplier participates significantly in the design of a part/component by executing detailed drawings based on rough sketches	29
Level 4: Supplier took full responsibility from concept to manufacture for the design of an entire part/component	0
Level 5: Supplier took full responsibility from concept to manufacture for the design of a system/subassembly that incorporated supplier designs	13
How frequently does your product group involve suppliers in the development process at the level indicated above?	
Less than 10 per cent of projects	13
Between 10 per cent and 30 per cent of projects	33
Between 30 per cent and 80 per cent of projects	38
More than 80 per cent of projects	17
Has your group developed and implemented a formal programme for the involvement of suppliers in the product development process?	
Yes: Programmatic ESI involvement	17
Currently developing/implementing a programme	21
No: ESI involvement is non-programmatic	63

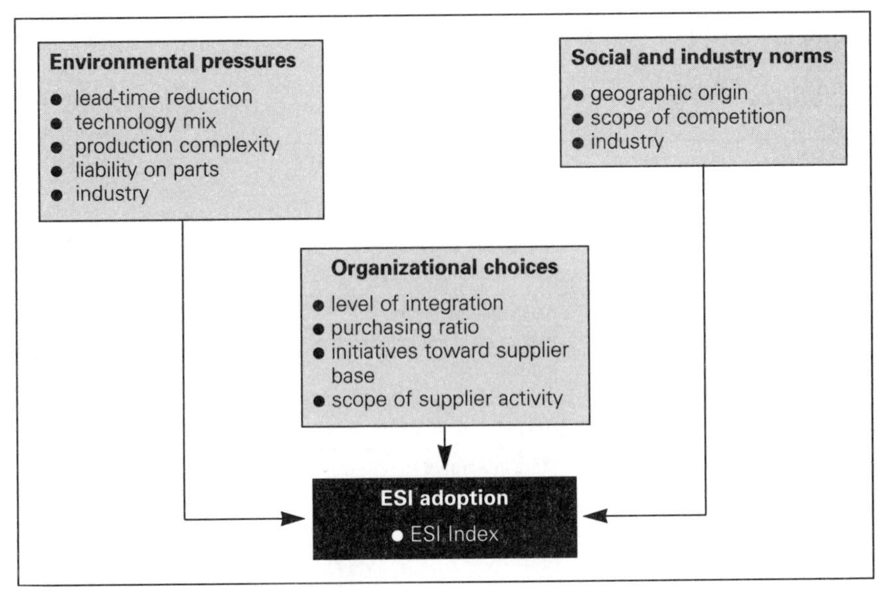

Fig. 4.5 The research model

score on the ESI Index. The responses underwent statistical treatment to assess the impact of each factor in a rigorous way. The outcomes are interesting but a word of caution: the sample is small – 22 companies – making the outcomes more exploratory than prescriptive.

Results

A total of 11 independent variables were tested for their effect on ESI adoption among the companies in our sample. Table 4.3 (overleaf) displays the results of analyses of variance (ANOVAS) across levels for each of these variables, in decreasing order of significance. Of the 11 variables tested, six achieved significance at .05 or better.

Of the three categories in the model that are presumed to influence the adoption of ESI, organizational choice received the most support from our results. Each of the factors in this category were significantly related to high levels of ESI adoption. Our results show that ESI adopters have developed a supply base that delivers more systems and subassemblies than other firms, and report more long-term cooperative agreements with a smaller supplier base. Environmental pressures were decidedly ambiguous. The level of production complexity showed little association with high levels of ESI and lead-time did not show a decrease

Table 4.3 The factors affecting ESI adoption

Variable	Category	Levels	ESI Index	Anova
Production complexity	Environmental Pressures	Fewer than 1000 units	22.2	F = 6.1913 sig. .0023
		1000 to 10 000 units	66.6	
		10 000 to 100 000 units	41.9	
		100 000 to 1 000 000 units	55.5	
		More than 1 000 000 units	30.5	
Supplier base initiatives	Organizational Choices	Accreditation programme	22.2	F = 3.9602 sig. .0178
		Reduction in the number of suppliers	42.2	
		Kanban	35.2	
		Delegation of quality control to suppliers	39.7	
		Long-term cooperation agreement	63.9	
Product integration	Organizational Choices	More integrated	32.2	F = 4.8353 sig. .0187
		About the same	47	
		Less integrated	66.6	
Regional origin	Social–Industry Norms	USA	51.1	F = 4.4559 sig. .0244
		Europe	32.2	
		Japan	41.6	
Industry sector	Environmental Pressure/ Social–Industry Norms	Electric Appliances	37.9	F = 4.017 sig. .039
		Consumer Electronics	29.4	
		Office Equipment	55.5	
Supplier scope	Organizational Choices	Suppliers have moved or are moving from providing parts/components to systems/subassemblies	48.1	F = 4.4545 sig. .0464
		They have not yet done so, or are not anticipated to do so in the future	35.2	
Proportion of purchased parts	Organizational Choices	Less than or equal to 70 per cent	36.5	F = 3.7601 sig. .0660
		More than 70 per cent	49.4	
Liability concerns	Environmental Pressures	Very concerned	45.4	F = .7395 sig. .4894
		Concerned	40	
		Unconcerned	33.3	
Lead-time reduction	Environmental Pressures	Less than 9 months reduction	48.2	F = .4419 sig. .6509
		10 to 18 months reduction	45.7	
		19 and more months reduction	37.1	
Mix of technology	Environmental Pressures	Growing in diversity	42.2	F = .1369 sig. .7149
		Stable	38.8	
Scope of competition	Social–Industry Norms	Local	44.4	F = .1965 sig. .8975
		National	44.4	
		Regional	33.3	
		International	41.9	

corresponding with increased supplier involvement. Assumptions that the mix and complexity of technology leads firms to adopt ESI were also not upheld, nor did we find support for the idea that liability concerns lead to ESI.

We did observe significant differences between industry sectors, however: the office equipment sector displayed a much higher level of ESI adoption than did electric appliances or consumer electronics. It is interesting to note that the office equipment industry does not produce (and consequently does not purchase) in volumes as high as consumer electronics and electric appliances – and smaller quantities of more sophisticated supplies are logically related to high levels of ESI. Norms in the office equipment sector also appear to be more open to cooperation with suppliers than those in electronics and appliances. We also obtained surprising results pertaining to geographic origin of the companies: the United States showed the highest level of ESI in our findings, ahead of Japan and Europe. This, of course, is contrary to experiences in the automobile industry. While consistent with our interview data and discussions with specialists, it is also true that the companies participating in our research were specifically chosen for their reputation as being 'advanced' in this domain and the likelihood of having developed progressive ESI relationships. Still, it is interesting that respondents in the US produced such a high ESI Index score, given the fact that a scant decade ago they were scrambling (as the business press characterized it) to mimic their Japanese counterparts. Our study fell outside the automotive context and provides empirical support for the common view that non-automotive companies across the globe are developing ESI relationships. The industries we studied differed, however: office equipment firms were significantly more advanced than the others.

Managerial Implications

The adoption of new buyer–supplier relationships to improve product development processes is increasingly found on the managerial agenda. Both our research and the existing body of knowledge point to several factors deserving of serious attention. First, it seems that the adoption of ESI is more a question of strategic priorities than irresistible external forces. This does not mean that managers should ignore the impacts of economy, competition, technology or culture; it simply suggests that ESI should be consistent with the internal preferences of a company. This was a consistent message in our executive interviews and in discussions with consultants and academic specialists. Situations do exist, however, where external circumstances should be heeded. Among the industry-specific factors, production complexity seems to be notable, as do the growing mix of technologies used in a product and the pressure toward shorter development lead-times.

There is a popular misconception that ESI is a pure 'purchasing technique', one that can be implemented with sufficient training. The opposite is true. The benefits of ESI can only be obtained when clear strategic priorities are set regarding make/buy or make/cooperate issues, and when an unambiguous policy has been established that guides managers in their transactions with the supplier-partner. This, in turn, can only be achieved when a company has selected preferred suppliers which it trusts, and has developed this supplier base in terms of the ESI process. Interestingly, the companies with the greatest ESI experience among those we studied (for example, Motorola, Xerox, Kodak) are also those that acknowledged the most difficulty with the implementation process and experienced the longest delays before reaping the full benefits. ESI requires the type of careful preparation that demands more time than companies unfamiliar with the approach tend to anticipate. Our counsel is to read and discuss widely, plan carefully and build flexibility into the organizational change process.

5 The Challenges of ESI

In the previous chapter we outlined the factors that lead to ESI adoption; in this one we discuss how to engineer an effective programme and manage the resulting partnership. Experience with ESI relationships is growing but, as we discussed in Chapter 4, the reviews are fairly diverse and there are no concise statements of how to manage the process. This is precisely our objective here: to put a roadmap in the hands of practising managers that will enable them to implement an effective ESI partnership. Our approach is practical and based on the life-cycle idea, that all ESI projects unfold over time: there are certain things managers can do to build a momentum for success and others that will almost certainly have the reverse effect.

Motorola Launches Its Initiative

Motorola has about 250 preferred suppliers in a pool of over 800. The 250 preferred suppliers are eligible for new product development programmes while the others furnish components for older and more standard products. Motorola knows a lot about its preferred suppliers – their capabilities, their quality, their error rates, their delivery histories. It works closely with them on component performance and together the partners work to shorten cycle times, improve quality and develop a larger pool of mutual benefits.

This was not always the case. Among the internal difficulties Motorola overcame to initiate ESI were objections from both engineering (which felt it could do the design better than suppliers) and purchasing (which did not want to work with suppliers without a quote). According to the manager we interviewed, '... our internal controls said you had to have three quotes before you could award a contract, so a whole sub-culture had to change'. The resources Motorola threw into the change effort were substantial. First, the ESI programme was fully and publicly supported by upper management. Second, the company published an ESI booklet which each General Manager sent to the product engineering organization with this message: we are now following these ESI guidelines. The General Manager of the mobile division, for example, stated flatly that a new design would not be issued if it had not been approved by one of the company's ESI people. Later, this same General Manager participated in a number of ESI orientation workshops for engineers. At one session, an engineer stood up and said, 'I've been designing chassis for 20 years and no (supplier) is going to design one better than me.' The General Manager leapt off his chair and replied, 'With that attitude, you won't be designing chassis for me in another week!' Harsh, perhaps, but Motorola had a serious need to improve quality and cycle times, so much so that it simply couldn't continue as it had been. This episode (and others of less dramatic impact) launched the idea of ESI and planted it firmly in the minds of Motorola employees – particularly engineering and purchasing employees.

'What got the momentum going were a couple of successes', according to another Motorola executive. Engineers began to understand that they no longer needed to spend a lot of time in administrative tasks, in supplier selection processes, and in manufacturing rework. The company promoted the initiative by giving awards to **both** the Motorola engineer **and** the supplier who developed a winning concept. Purchasers found themselves freed from the task of tracking design changes and suddenly, more able to devote their time to standard products. A group of 'Commodity Specialists' were named who followed a design and since these individuals came from purchasing, they had the authority to place initial purchase orders. 'In the long run,' said the executive we interviewed, 'ESI made the lives of engineering and purchasing a lot easier.'

Braun Steps Carefully

Braun produces a hair dryer with a spiky volume attachment. This attachment is made from two materials: a normal thermoplastic (the base and its spikes, both quite firm) and silicon (the tips on the spikes, which are soft and gentle on the scalp). Braun has years of experience with thermoplastics and a group of internal experts and moulding machines dedicated to this substance. Three years ago the company decided to soften the tips on this spiky attachment with silicon but it lacked the in-house expertise: Braun had never worked with silicon before. The company therefore researched the issue and after prolonged consideration, decided that it was better to 'buy' the expertise rather than make the component in-house. The next step was to find the best possible supplier. How did it proceed? By asking the plastic machinery suppliers it knew whom they considered the best silicon moulders. This led Braun to an Austrian supplier.

The two companies agreed that Braun would manufacture the thermoplastic portion and the supplier would concentrate on silicon tips. The result was what Braun considered a 'good partnership'. Now, however, Braun has brought the technology in-house to produce one version of the part in Ireland, while the original supplier retains production for a different version of the part in Austria. An 'exclusivity' agreement bans the supply of this part to any of Braun's competitors, though the supplier can use the technology elsewhere. Said our interviewee, 'If Braun had felt itself capable of doing this development in-house, we would have. The company does not like to go outside. This is the typical way that Braun collaborates with suppliers for custom-made parts.'

The experience of Motorola and Braun illustrate the variety of challenges facing companies embarking on the journey towards ESI. This is a longer and more delicate one than most converts would expect, one in which opposing winds can be tremendously strong and unstable. In the following pages, we would like to chart the terrritory by identifying the difficult issues confronting management as it starts implementing ESI and by suggesting some answers drawn from our field research.

Managing a successful ESI project is a balancing act: there are many steps along the way and a single slip may make it impossible to recover. Some manufacturers, for example, are set on achieving benefits for themselves at the supplier's expense. This leads to a control mentality in the place of a partnership attitude, and rather than investing in the project, the 'partner' supplier is encouraged to adhere rigidly to the terms of a contract. The end result is that the manufacturer 'wins' only if the supplier loses. Among the steps that can lead a company in this direction are showing mistrust in a partner and its capabilities, having a short-term attitude (the project, and the associated team, can be short term, but the attitude toward the relationship must not be), and lacking knowledge about the other's businesses and processes. A company that succeeds in ESI approaches things from a different point of view. The focus is on benefits for both parties from the very outset; projects of this nature have succeeded with real, and often substantial, benefits accruing on both sides. In the following pages, we will chart the steps that lead in this direction.

Managing the ESI Process

ESI requires a change in business-as-usual. In fact, the managers we interviewed all agreed that it requires a change of significant proportions. Developing an ESI partnership is an exercise in planned organizational change, an undertaking that requires care and forethought. But our interviewees also report that there is little mystery involved in the process: the basic requirement is a change process that orchestrates people and systems in a company. This, in one form or another, was a consistent message during our interviews with leading companies from around the world: successful ESI projects prepare carefully, execute a plan and monitor the resulting partnership. In what follows we will condense the relevant research knowledge and anecdotal evidence into four distinct phases: (1) the assessment of organizational readiness, (2) the development of internal resources, (3) the design of a partnership that fits both organizations and (4) the ongoing management of the partnership. Each of these phases involves a set of important issues which we catalogue in this chapter along with strategies for successful action.

Associated with these is the fact that highly successful ESI partnerships go a step beyond: they not only resolve the issues that arise but also undergird the effort with the values of trust and mutuality. These values are project-specific but they have their roots elsewhere – in the corporate culture. In case after case our research confirmed what is now emerging as a central element in successful partnerships: the alignment of systems, structures and thinking along an axis of trust and mutuality. These values are the decision-making touchstones that orient action and they should exist on both sides of a partnership, being as forthcoming from suppliers as they are from manufacturers. The pervasive importance of trust and mutuality will become clear as we review the conditions that lead to ESI effectiveness. These values are established in the first and support the remaining three phases of ESI adoption (Figure 5.1).

As we examine each of these phases we will begin by noting a set of objections that are commonly raised to an ESI initiative and then show how the successful companies that we studied have responded. We will

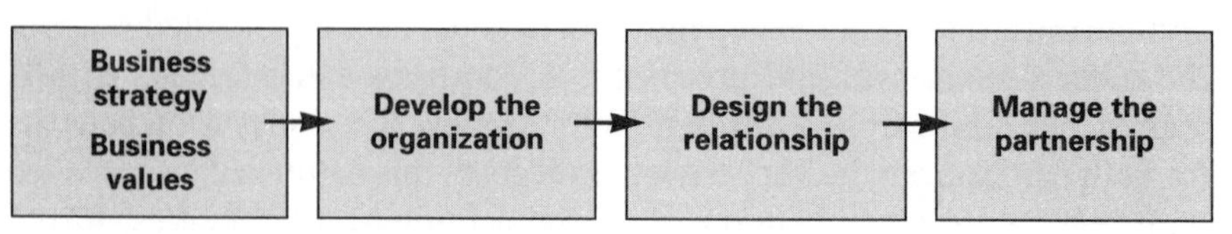

Fig. 5.1 Phases of ESI adoption

present the practical approach that these companies took to implement ESI and the logic behind it. The chapter is structured in this way – as a series of typical objections and successful responses over the life cycle – so as to provide the reader with an approach to managing ESI adoption that has practical value and academic backing.

Assessing Organizational Readiness

No organization can be effective at implementing ESI if it has not been prepared. Some companies are quite fortunate to be 'ready' for ESI because their evolution makes them compatible with the requirements of this new approach to product development. But, for the majority, it will be necessary to ensure that ESI fits with the strategy and the values of the organization. In some cases, these will have to be adjusted substantially, which obviously is not a minor effort. But without it, attempts at ESI are likely to stumble.

Business Strategy

If any message rang clear during our field research it was that ESI must fit with a company's infrastructure. This, of course, is a deceptively simple statement: any programme should 'fit' within an infrastructure. As will become clear in the pages that follow, however, ESI requires the type of large-scale assessment and development in an organization that some writers have termed a 're-engineering' effort. It therefore calls into question the fundamentals of a company's strategy, business values and

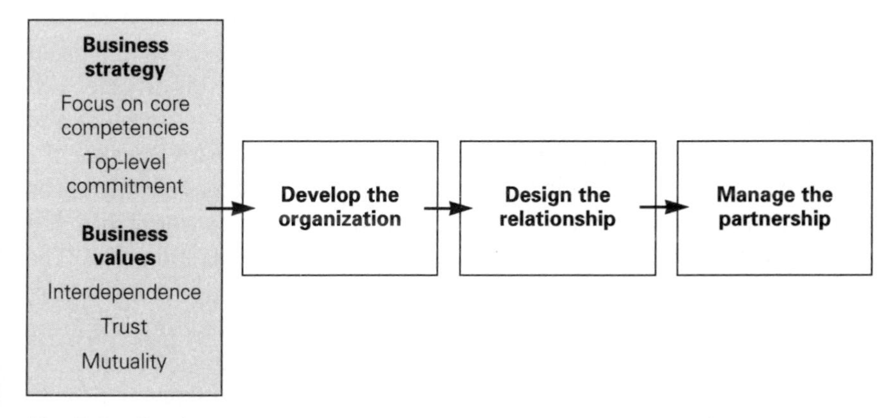

Fig. 5.2 Business strategy

operational procedures. When this magnitude of change becomes apparent, there are two objections that might arise:

Objection 1: *We will lose our distinctive competencies.*

The fear of losing organizational competencies was initially voiced by many of the companies we interviewed. This is consistent with any firm's reluctance to part with technical competencies that it has developed over a number of years. In some cases this objection originates in a company's engineering group, which misinterprets ESI as a downsizing move. In others, it comes from individuals who have built a career on technologies or procedures which ESI would supplant. Real issues are involved but the alignment of corporate competencies should not be confused with optimal staffing decisions which, of course, can be made to retain corporate talent.

Objection 2: *Maybe as an option but not as a policy ... people should decide when, where and how they want to use ESI!*

For political, operational or other reasons the argument may be raised that ESI should not be made a part of company policy. Instead, some will argue that it should be considered an option for development teams which would treat ESI like any other development tool. This thinking is seriously misinformed since it ignores the realities of ESI practice ... principally, that it is not a technique with an 'on again – off again' nature. It also poses the danger of seriously confusing staff and suppliers who will depend on consistency across projects and departments during an ESI development effort.

ESI requires the identification of core competencies. The firm that enters an ESI relationship is a firm that is choosing to focus on its core competencies. Whether this is an explicit decision and the outcome of carefully formulated policy, or an unintended result of the efficiencies ESI can bring, the net effect is that most firms move closer to defining and exercising their distinctive competencies. Why is this so? Because the reliance on expert partners in a product development process allows – even obliges – a manufacturer to concentrate internal resources where they add the greatest value. This stems from two dynamics: a push and a pull. The pull dynamic occurs when a company makes a decision to focus on core competencies and enlists ESI in the objective; it is 'pulled' into ESI as a strategic measure. The push develops when technological complexity, product proliferation and other development issues pressure a firm to concentrate its activities; it is 'pushed' into concentrating expertise on certain technologies and product components, and partnering to acquire others. Pure forms of the push and the pull are far less common than a mixture of the two. In fact, this was a strong message among the executives in our research: most ESI-adopting companies have decided they cannot do it all (push) and enlist suppliers in their product development cycle as a strategic measure (pull).

ESI is a matter of policy to be promoted by top management. Any programme of size and significance requires top management understanding and commitment and the same is true of ESI. We wish to state very clearly that in the majority of Western companies, ESI has required a level of organizational realignment that made top management support an imperative. This realignment typically involves roles, resources, functions, routines, schedules ... in short, the whole of an organization. Policy statements are therefore necessary and should be carefully developed by top management. At a minimum, the policy should include ESI objectives and rationale, the operating principles to be followed and a definition of the resulting interdependencies which capture – at a stroke – the internal and external realignments that will occur. The idea of interdependency was mirrored to us by a Motorola executive who had previously benchmarked practices at Sony of Japan. He had said to his Sony hosts that US engineers spend about 25 per cent of their time designing a component and 75 per cent understanding its manufacturing process. The Sony managers agreed that 25 per cent was about right for design work. But rather than spending the remainder of their time on internal problems, they said, they devoted it to understanding how to optimally integrate a supplier's development process in the Sony organization. Defining this type of interdependency will speak volumes.

After formulating its policy, top management should attend to three items. First, the ESI message should be sent both vertically and horizontally within the organization and steps should be taken to ensure that it is both received and understood. While this seems obvious, experience shows that it is anything but simple. Most communication programmes begin by applying the company's formal channels (memos, meetings, policy statements, newsletters, and so on) but as a second step, many companies have found it necessary to establish some form of training – or other ways of institutionalizing the mindset – to deepen the understanding. ESI is neither simple nor obvious and many organizations struggle to change their habits for years after having launched a programme. Second, top management should make it clear that ESI requires resources (time, money and personnel), that the necessary resources will be forthcoming, and that project planning will be based on factors that reflect this new reality. Lastly, after launching an ESI initiative, top management should remain both involved and detached ... that is, top management should exercise arm's-length oversight while granting the autonomy and organizational slack to project teams that allows them to experiment with novel ideas.

If top management involvement and guidance is necessary for ESI success, what happens in its absence? The likely outcome is a hamstrung project, wedged in a conflict between corporate and project-level strategies or in a nether land of competing values. Lexmark's Liberty printer suffered for such reasons: a lack of top management support at the project's inception made it difficult for the team to obtain human and other resources. The Liberty printer was therefore launched considerably later than necessary with a consequent diminution of market share. Other companies in our research were similarly hampered by indecision at the top and in several cases, existing supplier relationships were destroyed.

Business Values

Underlying a company's business strategy are its corporate values … preferred ways of dealing with internal and external 'realities'. These values are held in a collective corporate mind which (normally) is headed by top management. Explicitly or implicitly, the values expressed by top management have a normative effect so that individuals and systems tend to fall into conformance. ESI presupposes a set of corporate values to which some individuals may object. Two of the most common complaints are as follows:

Objection 3: *ESI creates unreasonable dependencies on (untrustworthy) suppliers.*

To the extent a company externalizes its programmes, functions and technologies it is likely that people will question the wisdom of the resulting dependencies. The issues are basic and important: Can the supplier deliver? Will the supplier be in business next year? How can we control quality, price and quantity? These are meaningful questions. Although ESI provides ways to address them it is likely that key personnel will be wary. In reality, however, the apparent dependency is an *inter*-dependency: ESI suppliers make significant investments in a development process on behalf of their clients. They become as dependent on the goodwill and consistency of a manufacturer as the manufacturer is on the supplier's competence and follow through.

Objection 4: *ESI goes against our traditions.*

An organization creates at least part of its identity by establishing certain ways of interacting with other organizations. Over time, these ways are lodged in history and habit and become traditions. Some companies consider their tradition of hard bargaining to be an intelligent response in the competitive environment: we are alone and succeed by dealing with outsiders unyieldingly. Furthermore, they often embrace the NIH syndrome – that which is *not invented here* has little value. This tends to spawn R&D and design/engineering functions that have strong domain interests in the company's product development processes: it is theirs and theirs alone. Objections to ESI will be raised to the degree that these functions perceive (and the operative term is *perceive*) that their domain is being appropriated by suppliers.

ESI assumes interdependence rather than dependence. ESI pivots on the idea of interdependence. Most companies have experienced being independent and competitive, or dependent and vulnerable, but few are accustomed to being interdependent. ESI, however, binds manufacturers and suppliers together in product development relationships which neither could undertake alone – certainly not as well, often not at all. Interdependence therefore becomes a strategic measure that makes company-to-company interfaces an operating framework. The ecology of business thus moves from being (perceived as) competitive to at least partially cooperative. Instead of fanciful philosophy, most of our executive interviewees labelled this idea hard-nosed realism. Consider this paraphrased excerpt from our interviews at Honda US:

> It's hard to get US suppliers to trust us initially but over time (4 to 12 years) they gradually understand what we're doing. They see the commitment Honda makes – some say Honda people are in their facilities three times more than all their other customers combined. This kind of interaction builds up trust, and the trust leads to the kind of business relationship we want. ... Suppliers that work in this way with us, who allow this kind of deep involvement, start making money and Honda asks for nothing in return other than commitment to the programme. You can't believe how excited suppliers can be to get involved with our development process.

Honda brought to its US manufacturing facilities a set of business values that it had developed over several decades in Japan, together with a conviction that US suppliers were capable of adapting. They were correct and the company is legendary for its innovations in the US automobile industry. On the other side of the Atlantic, a Finnish executive with Nokia expressed similar views from a more instrumental perspective:

> Paper guarantees for confidentiality don't work – they require too much paperwork, people don't do them, and so you don't have protection. Business morals and ethics are what's required. It's the business of working together that counts, it's the threat of discontinuing a mutually beneficial thing. That's the kind of business relationship that's needed when we do ESI.

ESI requires trust and mutuality on the part of both partners. The interdependence sought by Honda and Nokia made cooperation – not competition – the golden bough. Instead of disputing contract clauses and deadlines when a problem flared, they sought to work cooperatively toward a solution. We found that high-performing ESI companies consistently displayed a commitment to the values of *trust* and *mutuality*. Each partner assumed risks, each shouldered responsibilities and both believed they would benefit over the long term. An analysis of our interviews found the following trust-related themes recurring in one form or another:

- Trust versus scepticism
- Openness versus confidentiality
- Coordination versus fragmentation
- Cooperation rather than competition

In a word, ESI relationships pivot on values that differ from those found in the traditional Western business environment. Indeed, the term trust is becoming a central topic in all forms of interfirm collaboration and we will have more to say on this matter in the next chapters.

The concrete aspect of how trust and mutuality relate to the development of ESI partnerships involves the expectations and assumptions communicated by top management. Needless to say, an ESI project stands a poor chance of success if it is forced to operate in an antithetical environment. Values are held in a collective social space, which is to say corporate space, and they are transmitted through various communication media. People working in this social space take their cues and pattern for at least some of their behaviour on the values communicated. They form assumptions, in a word, and execute their tasks in ways they believe are expected of them: 'I know the top prides itself on being tough with suppliers so in this case, I'm going to be tough too.' Top management is a powerful communicator of such values, both explicitly and tacitly: its expectations concerning ESI will telegraph through a company via formal and informal channels.

Developing the Organization

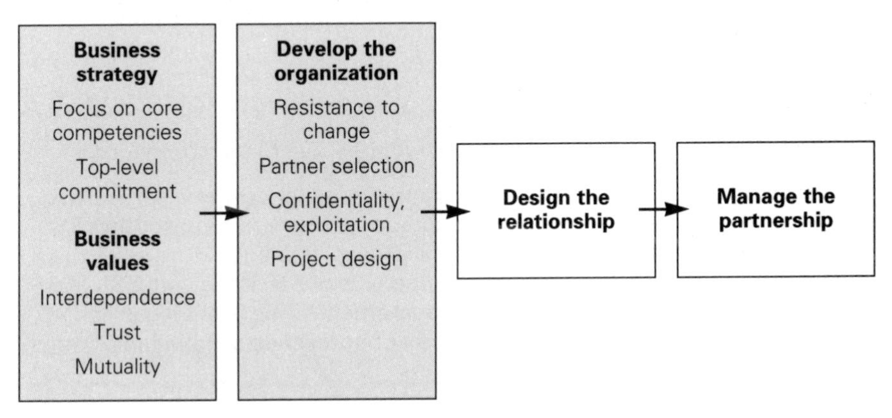

Fig. 5.3 Develop the organization

The strategy of focusing on core competencies, trust and mutuality highlights the fact that ESI is more than a simple set of procedures. It represents a new way of doing business that has ripple effects throughout an organization. The level of reform precipitated by ESI can be gauged by the fact that many of our interviewees referred to the level of adjustment as 'cultural'. The following is paraphrased from our interview with a Xerox executive:

> In the 1960s and 1970s we did our entire design in-house, for confidentiality reasons and so on. When Xerox got into CSI (continuous supplier involvement) in the early 1980s, it was a major culture change for the design engineers. In fact, even today, some design engineers are uncomfortable being totally open with a supplier – about the functionality of the part to be purchased, the product it's going to go into, when the product's going to be launched, the volumes it's going to be made in, and so on. There are still many people in Xerox who haven't fully made the cultural change. We need to do that.

Variations on this theme were expressed in most of the organizations we studied. In the typical ESI initiative there are three key issues that managers will encounter.

Objection 5: *Choosing a supplier is already difficult.*

The choice of a supplier is never an easy matter, no matter which purchasing approach is followed. With ESI, purchasing managers and individuals in other departments may fear that the task will become unbearable, akin to a lottery game. In reality, a new approach to supplier selection is being installed and old habits are often hard to change.

Objection 6: *Competitive advantage might be lost to the competition.*

It is natural for management to worry that product designs, particularly unique designs, remain within the company. Some will object that ESI puts these designs at risk. But a Skil-Bosch manager dismissed this concern in a sentence: 'Intellectual property is not an issue with ESI – get the supplier to sign a confidentiality agreement.' This is the legal side of the answer. Experience shows that other approaches are available.

Objection 7: *The supplier will never be a real team player!*

Building an effective team within an organization is a difficult task. People, departments, programmes, political clans and other forces often pursue different agendas. It therefore comes as no surprise that the idea of involving supplier representatives in a delicate product development project can sound the alarm: 'We have a hard enough time working among ourselves and now you want to involve an outsider?' The expectation is that suppliers will bring disruption rather than convergence into the team process. Others will argue that even if the composite team can be made to function adequately, the outsiders will protect their economic interests first and those of the manufacturer last.

Supplier selection must include new variables such as partnering attitude. Finding the right partner for an ESI initiative is an obvious imperative. The underlying assumption is that a manufacturer will locate and enlist a supplier that meets important criteria. What are these criteria? We found that manufacturers typically made judgements based on a combination of hard and soft criteria. Hard factors are those that can be objectively assessed while soft factors are subjective and rely on a judgement process. With regard to the first, four key issues regularly surfaced:

1. **Technical expertise** Is the supplier an industry leader? Does the supplier have technical depth that augments or complements the manufacturer's own? Does the supplier have design or engineering ability that will add specific value in the development project?
2. **Manufacturing capability** Will the supplier be able to grow with the project, technically, if the need arises? If the supplier has multiple contracts operating at a given time, will this compromise the project timetable?
3. **Logistics** Will the supplier be able to meet time schedules and delivery dates? Do lead-times have a good chance of being respected? Is the supplier located close to the manufacturer (most manufacturers prefer geographic proximity)? Will the supplier be able to respond to unscheduled requirements or emergencies?
4. **Cost** All other criteria considered, is the supplier within the manufacturer's cost targets?

While soft criteria are more ambiguous it is noteworthy that executives participating in our research treated them with at least as much respect as the hard factors, and frequently more. They spoke of the following as cornerstones in a partnership:

5. **Communication** Does the supplier evidence the willingness, and the ability, to communicate well, consistently and candidly? Is the supplier capable of maintaining confidentiality?
6. **Commitment** Is the supplier's top management committed to ESI? Have they developed their operation to the point of having worked in this mode before? Does the supplier share the manufacturer's sense of urgency?
7. **Cooperation** Does the supplier show a willingness to be open and deeply involved with the manufacturer's ESI process? Does prior history indicate that the supplier will be reliable and trustworthy? Does the supplier have the ability to learn during a development effort and capitalize on new knowledge?

These seven criteria generate an imposing checklist. Not every supplier will pass the test, nor should they. The fit between a manufacturer and a supplier is partially project-specific, partially corporate and always unique.

Overall, however, it seems a safe bet that the supplier who meets these seven criteria will form a suitable partner. Once chosen, it will be important to involve that company's representative fully in the development effort. ESI manufacturers treat their qualified suppliers as important manufacturing assets and deploy them accordingly. On this point, for example, Motorola administers a 'supplier advisory board' which consists of 15 rotating suppliers with whom Motorola holds meetings three times a year. The agenda? A review of the working relationship; an examination of problems; a series of suggestions for improving the future partnership and a serious discussion of how Motorola and the partner can deepen their involvement.

Losing competitive advantage through leakage is uncommon. The idea of involving a supplier in the deep recesses of a manufacturer's product development programme will give rise to confidentiality and exploitation issues. This is to be expected. With regard to confidentiality, the concerns revolve around intellectual property rights for innovations produced with or by a supplier under the terms of an ESI agreement. Closely related is the exploitation issue of how the benefits from jointly developed products and/or innovations will be shared – a significant matter for suppliers who plan to invest considerable R&D to develop a product. To what extent should they profit from this investment?

Both of these are potentially contentious issues. Surprisingly, the vast majority of manufacturers and suppliers in our research reported very few examples of innovation leakage. They all urged certain precautions which, in general, involve confidentiality and exploitation agreements that are developed from the start. We will detail these in an upcoming section and here wish only to emphasize that management should keep the issue clearly in mind and expect to develop an agreement that protects its interests. Not doing so can be dangerous: the German appliance manufacturer Braun, for example, lost the benefits of an important innovation when a key supplier shared co-development information with its competitors. As a result, there are now several manufacturers who have the very same ('innovative') coffee pot handle. Braun not only severed the supply relationship, it remains an ESI sceptic. Xerox, by contrast, has made it a policy to develop and sign a confidentiality agreement at the outset of a project … but then puts it away. Xerox management believes that if the agreement is referred to during the life of a project, the partnership is in trouble.

Supplier involvement requires appropriate leadership to overcome team-building issues. Together with the systems and structures that support them, cross–functional teams are often at the heart of an ESI programme.[96] Broad representation on this team is indispensable and early identification of the members is a key step. Those who participate on a cross-functional team should possess the ability to leave traditional thinking behind and mix their functional loyalties with others on the project. This is not always easy. Most professionals are steeped in a background that conditions their outlook and politically, their functions often compete for resources and protect the integrity of their boundaries within a company.

Our own research found that two groups in particular – design/engineering and purchasing – have this tendency. Those with seniority in these groups may react to ESI as a challenge to their traditional assumptions and ways of working. It is not uncommon for purchasers to disagree with procedures that fail to pit one supplier against another, and for some engineers to believe that management has lost faith in their technical abilities. The task is to make these groups and individuals feel responsible for the implementation of ESI and this is best accomplished in a team context.

Largely through this cross-functional team, leadership is crucial to the project's success. If the project leader is not powerful, communicative, ambassadorial, visionary and adept at internal team management, the project may well be in jeopardy. This is an important role that requires competence and talent. The leader, for example, should be able to dissolve functional allegiances and lubricate cross-functional relationships; should act as a 'linking pin' within the organization to ensure that impressions are managed and resources (time, money, personnel) keep flowing; should be able to consolidate diverse views and interests in order to keep the project focused and on track. Research also shows that ESI projects are more likely to succeed if the leader has significant power and prestige inside the organization. Project leadership is a key issue and considerable care should be exercised when selecting the person for this role.

Lastly, the communication structures within and around an ESI project are strongly related to its success. Internally, the variety of information provided by members with diverse functional backgrounds is important, but this information must be channelled into effective problem-solving methods which, in general, are a function of the team's sophistication and the competence of its leadership. With regard to external linkages there is evidence that the more a team gathers and incorporates information from outside the project, the more successful the development effort. Furthermore, it is inadvisable to rely on a given function to manage the information germane to its expertise; rather, it is better to have multiple inputs from various members of the team channelled into the cross-functional decision process.

These points relate to suppliers as well as manufacturers. Kodak, for instance, believes that suppliers should have a position in the ESI project that is equal to its own departments and employees. Its ESI supplier representatives are not junior partners that feel invited by Kodak to contribute only when asked. They are full team members who develop a cognitive and emotional bond with the project, just as Kodak employees do. 'We want to have them on the project team photograph', explained a Kodak executive.

To summarize, it is important to ensure that strategy and values are properly aligned, that top management is committed and that organizational resources are forthcoming. The managers initiating ESI should understand that it tends to redistribute power away from functional domains and hierarchies and concentrate it in an integrative project. Working within traditional boundaries is anathema to ESI: roles, responsibilities and routines will be realigned and employees schooled in older traditions may show resistance. Cross-functional project teams have been an effective response to this issue and in many cases, become the entity that both symbolizes and concretizes the principles of ESI.

Design the Relationship

Once a manufacturer and a supplier have agreed to work together the new partners are faced with three issues: definition, integration and execution. Their first task is to define a suitable basis for working together, the second is to develop integrative methods for collaboration, and the third is to manage the project such that both companies reach their goals. Together,

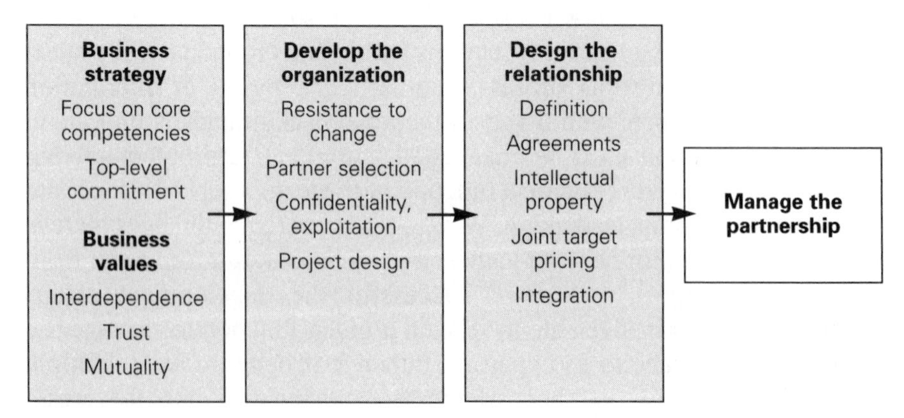

Fig. 5.4 Design the relationship

these three objectives formalize the interdependency we discussed earlier. There are a number of ways to manage this interdependency successfully and in this section, we treat the issues of definition and integration. In the last part of this chapter we will focus on executing the design so that the ESI partners reach their objectives.

The Definition of the Partnership

What we term 'definition' occurs at the outset of a partnership. As a first order of business the partners should specify four issues that are sure to loom large as the working relationship gets under way: product development, supply, confidentiality and price. These are much like the tips of an iceberg since they arise from economic and performance concerns that both partners are certain to have. In case after case our research confirmed that successful partnerships set expectations in each of these areas early on. Some companies were detailed and meticulous, others less so, but all addressed the four issues as a first step. The resulting agreements often defined procedures as well as outcomes. Both are important and it is a mistake to focus solely on the outcomes desired. The goal is to assure both partners that their interests are being protected. Three objections may arise in this regard:

Objection 8: *ESI will require a complicated agreement.*

Managers are often anxious about the contractual implications of ESI. Because suppliers will be involved in design activities that entail a level of secrecy, and because the commitment to the source of supply is expected to be long term, managers may expect that the contract will be sophisticated, expensive and nerve-racking to negotiate. Fortunately, this is often not the case since most ESI agreements are built on a history of trustful relations.

Objection 9: I*t will be hard to control our intellectual capital.*

Despite the research evidence and numerous experiences to the contrary, the risk of having a supplier share jointly developed technological innovations with a manufacturer's competitors is likely to be voiced. True, this is a legitimate business concern and it would be foolish to ignore the possibility. But there are also effective ways to handle the matter.

Objection 10: *Suppliers may try to take advantage of us through excessive prices.*

The fear of being 'locked-in' with an ESI supplier is sometimes expressed because the supplier is placed in a monopoly – and therefore powerful – position. Again, this is a reasonable business concern but our research shows that it is not a significant problem and wider reports of misbehaviour in the industry are rare. The pricing issue remains a serious one, however, and our research disclosed several ways to address it.

Most ESI agreements are straightforward. ESI agreements typically make a distinction between technical development and product manufacturing. The development agreement is usually an informal document that specifies two expectations: the level of the supplier's contribution (feedback versus actual design work, for example), and the nature of the supplier's involvement (reviews, prototyping, texting, pilot series). The development agreement does not deal with confidentiality issues: as we will see next, it is generally produced at the beginning of a relationship with the intent of setting expectations and ensuring smooth and coordinated work.

The supply agreement is typically more formal than the development agreement and specifies quantity, delivery and pricing. It rarely spans the entire 'part life' cycle of a component, typically being limited to an annual time frame for which volume as well as price and price reduction schedules are specified. Some companies simply exchange letters at the beginning of the year, allowing actual orders to dictate the supply arrangement. Most, however, fall outside this trust-based framework and develop a formal document.

We are not saying that detailed and precise contracts are to be banned. But, experience is showing that, generally, ESI relationships do not require long and complex documents that only large companies could afford.

There are several ways to control proprietary information. Confidentiality agreements between manufacturers and suppliers fall into one of two categories: formal and informal. Formal agreements are signed between the partners to govern the use of innovations produced in the partnership. Most limit both the supplier's and the manufacturer's ability to divulge proprietary information and prohibit the transfer of technology or information to third parties. There are also manufacturers and suppliers that operate on an informal basis: a principle of non-disclosure guides the ESI project and the understanding is that innovations will neither be used by the supplier with other clients, nor by the client with other suppliers (in the hope of bettering price conditions). Agreements such as these are generally formed on the basis of long relationship and familiarity between the partners. Two factors are generally in play. One is the trust and ethics established between companies and individuals over a period of time. The other involves the economics of a long-term supply relationship and the establishment of a reputation of 'honesty', which many suppliers find advantageous.

With regard to patents and the exploitation of products developed in a collaboration, most manufacturers and suppliers adopt one of three ways to address the matter. The first is an agreement that all innovations and products developed in the course of a partnership will remain the property of the manufacturer. This is a proprietary approach which most manufacturers expect but which, in practice, proves difficult to enforce. The second is an agreement which makes products and innovations developed in a partnership available to both parties for exploitation. The terms of this benefit sharing are, obviously, unique to each project and a matter for negotiation. The third approach might be termed shared-after-a-time and is based on an agreement that products and innovations developed in a partnership will remain the property of the manufacturer for a period of time, and subsequently be made available to the supplier for exploitation. This is the most widely practised accord. It is fair to the supplier in that it acknowledges contributions by giving title to the innovations, and it protects the manufacturer from early emulation. What time lag do companies typically adopt in this case? Some fraction of the life cycle (20 to 33 per cent) is an order of magnitude that some experienced companies would recommend.

Joint target pricing to address the issue fairly. Pricing is a sensitive issue because it deals with baseline economics. Typically, the manufacturer is concerned that given the absence of competition, the supplier will have little incentive to offer low prices. Suppliers, on the other hand, often fear that their unique and costly investments will commit them to an unbalanced relationship where the manufacturer is able to exercise undue power. The large plastics moulding firm Nypro, for example, considers that many manufacturers for which they worked intially expected to keep all margin increases resulting from Nypro's design improvements for themselves which, obviously, as a supplier, they resented. Now, Nypro has become very selective and only works with clients that have a more open-minded attitude about pricing. Xerox, on the other hand, admits that sometimes as a client they might become overly demanding, for instance when their designers come up with excessive change orders and tend to be oblivious to the cost implications of their requests.

The traditional response to this dispute about pricing is a spiral of bargaining and counter-bargaining which is aimed at mitigating economic surprises. A 'fair' pricing scheme begins by countering these negative postures. It does so by opening the books: each partner informs the other of the cost structures involved. Openly and honestly, each partner shares this information with the goal of increasing the benefits available to both.

This is not an easy prescription. But despite the fact that an experience curve is involved for both parties, many companies have applied it with success. Honda of America, a firm with considerable ESI background, suggests a policy to share equally the cost savings resulting from collaboratively generated improvements, thus reinforcing the partnership principle with its suppliers.

Although these pricing agreements are set in writing they are often living documents, open to examination and change. As a Kodak executive explained, 'It used to be a bartering game with suppliers: how much can they make, how much can we make. We didn't understand each other … that we need to co-exist, to break down the barriers and understand each other. We did this by a lot of talking, a lot of travel, and a lot of trust.'

Based on these types of industry reports and the results of this and other research programmes, our conclusion is that the pricing issue is best handled by thoroughly communicating the respective cost constraints and revenue objectives with one's partner. A significant level of transparency coupled with a genuine desire to obtain mutual and fair benefits are the two foundations of a sustainable agreement. These can be achieved through 'joint target costing'.

Joint target costing is an approach to setting the cost to be achieved for a part, given the overall price that the end-user is willing to pay for the product. The cost target is set to guide the supplier in its development and this is typically done by the manufacturer alone. The supplier then works with this cost target to develop a supply proposal and discussions between the partner bring the matter to a mutually satisfactory conclusion. Advanced ESI implementers, however, actually involve the supplier in developing its cost estimate. By doing so, they establish a target that is based on deep knowledge of the development constraints involved – assuming, of course, that the partners' ESI relationship is indeed 'advanced'. Target costs set in this way may actually be lower than projected by a manufacturer, as Renault experienced in the development of electrical wiring for its Twingo. Conversely, if the target cost becomes higher than projected, it might be for very good reasons, for example component longevity. In either case, the ESI relationship is jeopardized if the target cost is a fiction designed to gain advantage for either partner. This again emphasizes the necessity of building a proper relationship between partners from the very start.

The Integration of the Partnership

An ESI partnership becomes most authentic when the two parties develop a cognitive and emotional commitment to the project and share in its identity. This is the full sense of the term 'integration' whose implications, of course, may create considerable anxiety.

Objection 11: *Without competition between suppliers, we'll be taken for a ride.*

Assuming 'fair' pricing targets are initially agreed on, some managers (particularly those with a purchasing background) may argue that the supplier will not be vigilant for lack of competitive pressure. Their objection will be that over time, the supplier will relax and allow costs to drift upwards.

Objection 12: *The supplier won't be able to follow our pace.*

Technological excellence is never achieved once and for all; rather, one reaches a level of technological competence that is constantly challenged as new technology comes on stream. Managers may argue that, 'Even if this supplier is on a par with our in-house capabilities today, how can we be sure they'll be good in a couple of years – when the technological landscape has changed? Will they be able to follow our pace? Will they keep up with the industry?'

BOSE initiated what it calls JIT-II in 1986; seven years later its achievements were featured in *Industry Week*. JIT-II is an advanced system where supplier 'in-plants' are physically housed in BOSE facilities and become integral parts of the BOSE organization. As one manager said, 'JIT-II is so catalytic it's unbelievable. No more salesmen calling and buyers buying. The in-plant places orders himself in conjunction with our MRP system and then practises concurrent engineering in-house on a full-time basis.' In 1995 BOSE had 12 in-plants from nine suppliers on its premises.

These people were sourcing materials, fighting for lower component prices, managing logistics and in some cases even coordinating activity in the corporate office. How deeply involved were they? One BOSE representative explained that in-plants actually purchased materials for BOSE from their own companies – and often in competition with other vendors. When BOSE checked pricing in the market, '... 99 per cent of the time, the in-plant has had the cheaper price. This is due to inside knowledge.'

BOSE has gone a step beyond the cross-functional project team; it is reaching toward a cross-organizational model of ESI management. Blending the work of two companies on this scale requires that structures, systems and methods are all brought into a high degree of alignment. The anchor, the unmovable point of reference, is the project itself and an important factor is the structure of communication and coordination. As a Lexmark executive told us,

> A good partnership takes a lot of talking. What goes into a good partnership? The same thing that goes into a good marriage or a good friendship: communication and lots of it. It's like the time we thought we had an agreement with one company to start tool production while the drawings were being finished up but despite meetings to this effect, we finally had to take it all the way to their Vice President to get things understood. Accuracy of information is critical, especially where ESI is new for one or both of the parties. It's a problem of abstract terms that we all understand in our own way.

The practical side of ESI is a world of interpersonal communication and interlocking schedules; managements should design their project structures accordingly. Common mechanisms include on-site supplier representatives, a geographically close supplier location, designated liaisons from both companies, a generous travel budget, tele-conferencing, shared CAD files and regularly scheduled meetings. As a Diebold manager told us,

> The problem is, everyone's in a hurry and there's sometimes a lack of discipline about taking the time to agree on schedules up-front. This is critical ... engineers are not business people and they don't naturally stop to ask how much this increase in tolerance might cost, how much longer it will take to produce, ...

Another Kodak manager told us that ESI's two essential elements are, '... an empowered team and a strong product leader who holds all the pieces of the puzzle and has the matrix on the commercialization of the product'.

ESI does not necessarily mean absence of competition. Most purchasing managers (and most organizations) are accustomed to pitting one supplier against another in order to win the lowest price. ESI operates on a different logic: a chosen supplier is presented with inside knowledge and asked to proceed on a basis of fairness and this, understandably, requires a new mindset in the purchasing function. The mission becomes *supply base management* – which starts with a clear understanding of the potential suppliers for a given type of parts/subassembly and their competitive postures.

Fuji Xerox, the company with the most impressive ESI experience in our research, goes several steps further. It proved that some degree of competitive pressure can be maintained during the ESI relationship. In certain circumstances, when cost reductions are critical for certain supplies, Fuji Xerox involves two subcontractors in the design activities. Both contribute to the development and both will receive orders when the project moves to the manufacturing stage. At this point, however, there will be a degree of competition between the two since their Fuji Xerox 'market share' will be affected by their price performance (there are guarantees that a supplier will not be eliminated outright). Needless to say, this requires a high level of confidence with and among the suppliers … something which can only be achieved through a history of trust and mutuality.

Suppliers will follow the pace if they are supported. At one point in time a group of purchasers at Bosch Power Tool were arguing for the introduction of ESI but running headlong into a problem: a number of engineers simply failed to see how small suppliers could possibly provide a company like Bosch, a clear leader in its industry, with competence they did not already possess in-house. This objection is sometimes voiced as ESI gathers momentum: how can a supplier possibly match the quality and capabilities of our own organization, today or tomorrow? It is sometimes coupled to a prejudice against small suppliers, the thinking being that size runs in parallel with competence. In most scenarios, however, and in keeping with the principles on which ESI is founded, the suppliers selected for a project already have specific competencies that exceed those of the manufacturer.

Furthermore, provided the opportunity to progress through involvement in challenging development activities, most suppliers will readily 'follow the pace'. A large body of research has shown that innovativeness and the quality of client relationships are positively correlated. MIT's Eric Von Hippel, for example, has shown that clients are the main source of innovation in many industries. Those that are open to their suppliers in this way often benefit from innovations which would otherwise have gone undeveloped. The reverse process is equally clear: if a manufacturer adopts a protective attitude and refuses to expose suppliers to important development activities, the suppliers will achieve stasis unless competitors place a challenge before them.

Manage the Partnership

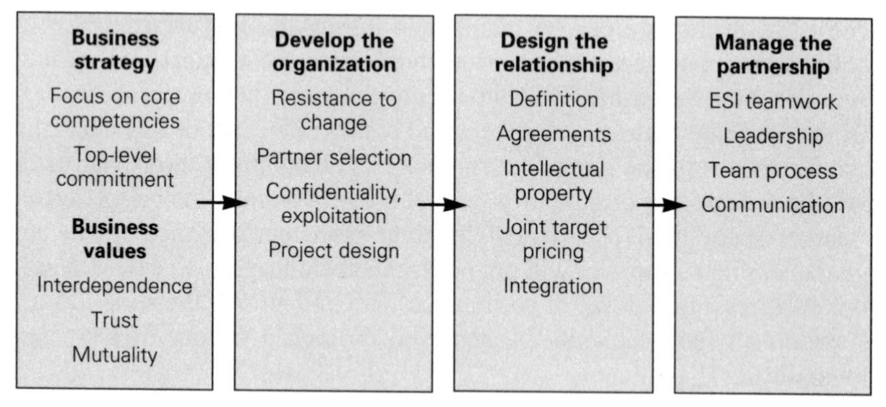

Fig. 5.5 Manage the partnership

Once an ESI project is launched the partners are faced with a number of management issues. While there are certainly other factors that affect the eventual outcome (product concept and design, market factors and consumers' evaluation of quality, innovation and price, and so on), four issues are consistently cited as important elements of a successful effort: leadership, partner relations, team process and communication patterns. The astute reader has noticed that, with few exceptions, the recommendations and observations in this chapter have dealt with the process requirements of an ESI project rather than its content. The management of an ESI partnership should follow this example. The goal is to create and maintain an environment where skilled experts can apply their talents with as few barriers to creativity and execution as possible. This is accomplished not by telling a design engineer what to do, for example, but rather by creating work structures where (s)he can discover the best way forward. Four objections may nonetheless arise.

Objection 13: *Our staff won't be able to change their old habits.*

ESI challenges many existing work customs. Some managers may argue that while they are willing to change, others in the organization will not be able to evolve. If this is the case, ESI becomes an expensive trial that has little chance of success. This objection, unfortunately, is sometimes valid. The answer, however, is not to succumb to the reality but rather to change it.

Objection 14: *With outsiders involved, the project team will become unmanageable.*

ESI requires that supplier representatives become full members of the manufacturer's product development team. This sometimes generates a negative reaction on the part of managers who believe that the heterogeneity of employers will create divided loyalties.

Objection 15: *These supplier people simply don't understand how we work.*

A version of the previous objection, this argument is that involving outsiders in daily work tasks will be difficult at best. From another perspective, however, it is possible to say, 'We simply don't understand the way our suppliers work.' ESI calls for considerable knowledge exchange, not simply with regard to the technology but also in terms of work processes.

Objection 16: *ESI is nice but it takes too much time.*

Once the merits of ESI are sorted out, diehard opponents may admit that the approach makes sense but object that despite its advantages, ESI is a luxury that requires significant installation time precisely when the top priority is time-to-market.

To promote a change of attitude, leadership must consistently reiterate priorities. An ESI project requires two types of leadership: one related to daily management and another to the informed control exercised by top management. In most cases the project leader occupies a pivotal position in both regards: (s)he must create bridges to upper management, effective links with outside organizations and furnish the project with resources while guiding its progress. Project leaders that have significant decision-making power within the organization are generally more effective than those who do not. Why? Because they usually have access to the necessary resources as well as the ears of top decision makers. Leaders with this command of political dynamics tend to be more effective than others.

Related to this is the matter of vision, the leader's ability to synthesize chaotic information into a clear and coherent view of the way forward which creates a structure for work in the project. The leader must be able to specify needs with designers, researchers and engineers, solicit the input of expert outsiders, the insights of purchasers and manufacturers, keep market research clearly in view and mesh all this together into a clear view of the product and how to produce it. Clearly, a difficult task involving considerable maturity and skill.

A less obvious form of leadership relates to senior managers who provide a form of *subtle control* that leads to superior development processes and ultimate product performance.[99] This concept was coined to describe an aspect of Japanese ESI relationships and is defined as '... the vision necessary to develop and communicate a distinctive, coherent product concept'.[100] Top management maintains an arm's-length relationship with the project, setting strategic priorities, but providing the autonomy to reach them however the project determines best. It has been argued that this is a critical element in developing creative products.

Appropriate team work is actually effective to change people perspectives. Teams are not built once and for all – they have to be maintained and managed to ensure cohesion and effectiveness. The practical recommendations for managing teams of this nature generally stem from research in small group dynamics which, we find, is valid in the ESI context with one contingency: ESI teams are likely to have a membership that is both varied and changing. A core development group tends to remain stable while other members rotate in and out as the needs present themselves. The net effect is that of re-establishing certain developmental milestones that a more stable group would pass and leave behind.

With this caveat, it is true that all groups move through stages of development and increase their effectiveness with maturity. Interested readers are advised to pursue the details elsewhere, but as an example of the actions that can lead to these qualities consider the following. The way members are recruited to the project presents a developmental issue at the very first. Whether they self-select into the project or feel coerced by senior management will have an effect on later performance. During its first meetings a team will usually test for structure – people will question the objectives and their elasticity, the nature of relationships, the types of behaviour considered acceptable and so on. This typically leads to a stage of conflict where power (who will have it and who won't) is the underlying issue. If the leader manages these issue effectively, the conflict will subside or be put to constructive use but in any case, a sense of the power hierarchy will develop. Once these and other behavioural norms are established, most groups begin their serious work and the ideal leadership strategy is to facilitate work interactions and an open exchange of views. Mature groups thereby become committed and members settle into various roles: among the more prominent are leadership roles, task roles that focus on goal accomplishment, maintenance roles that nurture the group's social processes, boundary-spanning roles that extract information from inside or outside the group and coordination roles that catalogue and synthesize work progress.

This sample of issues points to the fact that consciously or not, high-performing ESI teams successfully negotiate a series of challenges. If the team has a clear sense of purpose and work process, an understanding that suppliers and others will rotate in and out of the membership, the necessary social-psychological infrastructure and resources and a leader attuned to group dynamics, the project is better positioned for success. If any of these are lacking, breakdowns and hesitations are more likely.

In case after case our research confirmed that team members need to keep the project and its goals clearly in mind as they move forward. The simple fact is that changing behaviour is difficult: despite words to the contrary, some manufacturers find it difficult to relinquish control and some suppliers are ill at ease when working in tandem with a manufacturer. Olivetti-Canon, for example, tends to delay final supplier selection even though the supplier involved in the design usually gets the job. Manufacturers have also been known to hinder a supplier's development with numerous engineering changes and suppliers are sometimes reluctant to explain production details for confidentiality reasons, hesitate to be honest for fear of losing the contract, or overestimate capabilities in order to get the business.

Difficulties are especially likely to arise if work processes or time requirements are poorly understood. The usual prescription that development and delivery schedules should be made explicit turns out to be feeble advice in reality. Schedules are necessary but the partners must go much further to ensure a smooth operation – notably, they should strive to understand and accommodate the work processes in each company.

Delays and breakdowns in partnerships often result from little appreciation of the way different companies accomplish their work. A touch of this parochialism created design delays at Lexmark and tooling problems at Minco, for example. Neither company had a clear view of methods and expectations in the other and a series of confusing signals led to a round of crisis meetings ... after which the project moved forward on a firmer footing. Diebold executives have learned to circumvent this problem by having suppliers participate in project scheduling; Bose and Kodak both recommend engineering and purchasing 'in-plants' from the supplier to learn about the manufacturer's business and product development process. In short, each partner needs to thoroughly understand the product development process of the other.

Communication promotes a partnership attitude.

> Communicate much, frequently and openly. You need to have effective communication and to work as a team so you don't miss opportunities. You [the manufacturer and supplier] need to understand shared destiny, that we need each other to survive.

This advice was offered to us by a Xerox executive. The research and experience in this area agrees that communication is a common concern at two levels: project team and organization-to-organization. Focusing on the first, it is clear that certain dynamics enhance the flow of communication in a team and that others hinder it. A widely respected guide for structuring effective interpersonal and group communication is adapted to ESI in Table 5.1.[102]

The communication patterns on the left of this table are those that provide a solid foundation for cross-functional teamwork, particularly in view of its rotating membership. These communication patterns are also related to the management of an effective team and are frequently a part of ESI team-building exercises. Cross-functional project teams must deal with the fact that professionals with different functional backgrounds have markedly different perspectives which often lead to difficulties when working together. Successful cross-functional teams have a concrete and iterative communication strategy where team members focus on a specific problem and discuss the issue at length – in the style above – until a resolution unfolds.[103] It also seems that a certain disrespect of functional divisions is helpful: for example, not allowing individuals from one functional area to dominate the topic when it deals with their expertise.

Apart from team communication it is important that the ESI project

Table 5.1 Effective and ineffective communication patterns

Effective communication patterns	*Ineffective communication patterns*
Description: giving and asking for information	Evaluation: praising, blaming, passing judgement, calling for different behaviour
Problem orientation: jointly collaborating in defining problems and seeking solutions	Control: attempting to persuade others by imposing your personal attitudes on them
Spontaneity: dealing with others honestly and without deception	Strategy: manipulating others
Empathy: identifying with others' position or problems	Neutrality: showing lack of concern for others
Equality: de-emphasizing status and power differences, respecting others	Superiority: reflecting dominance over others
Provisionalism: postponing taking sides, being open to new information and interpretations	Certainty: being dogmatic, wanting to win rather than solve a problem

relates effectively with its parent organizations. This has two aspects: one political, the other informational. Politically, the team needs to have effective links with outside groups for resource inputs. This means interacting effectively with senior management for resource allocation as well as with external suppliers to involve them fully in the project. But the team requires information as well, and the richness of this content is a factor in the quality and success of the eventual product. The evidence indicates that highly successful project teams use a strategy of scanning for a wide variety of information, importing it into the team, evaluating it for usefulness and incorporating what fits into the work. The more comprehensive and extensive this scanning activity is, the more successful the project.[104]

In terms of organization-level communication between a manufacturer and supplier, the quality and frequency of communication between the partners is positively related to the project's effectiveness. In practice, this argues for a process where each becomes familiar with the other and the prescriptions are classic: extensive use of the cross-functional team, the use of 'in-plants', regularly scheduled review meetings, and frequent use of the telephone, the fax and the airlines. Many ESI partnerships make use of advanced technology to facilitate the day-to-day operation of a project. Lexmark, for example, electronically transfers solid model drawings to plastics suppliers who have a technologically compatible machining centre, allowing tools to be developed in a matter of hours. Diebold relies on the same method and in one case (an ATM panel) had parts starting into production before the drawings specified all the final dimensions. Another supplier put a CAD terminal in the middle of a GE design team and cut the project's cycle time dramatically.

The time spent on ESI installation is recouped in later phases. Time invested in detailing a new product design during the beginning of a project can be demanding in a collaboration and may therefore seem unnecessarily long and cumbersome for those accustomed to the (possible) efficiencies of an internal process. This has to be admitted. The experience with competent ESI projects, however, is that this investment pays handsome dividends over the rest of the development cycle.

As we saw in Chapter 4, the vast majority of firms which have implemented ESI claim that it speeded up their development cycle. There is also evidence that projects incorporating ESI activities are among the fastest the manufacturers have ever launched. One example, among others, is the development of the Triathlon injection cleaner by Philips Floor Care, the Philips company which manufactures vacuum cleaners. This new appliance incorporated more technology than most other products made by Philips Floor Care as it was designed to accomplish three functions – dust cleaning, water pumping and shampooing; other machines in the product line accomplished only two at most. Yet, through an ESI process with considerable supplier participation, the Triathlon was developed in 18 months versus the 30 to 36 months to which this business unit was accustomed.

Framework for Success

As we noted at the outset, the balancing act that firms will perform in managing a successful ESI project is indeed a precarious one. There are many steps in the process and many opportunities to fail. On the other hand, the potential for reinventing processes and augmenting the balance sheet is substantial. ESI requires a change in business-as-usual and developing an ESI partnership is an exercise in planned organizational change. Companies that prepare ESI carefully, execute a plan and monitor the resulting partnership are those which are most likely to succeed.

In this chapter we have discussed a number of key issues within a four-phase framework – the assessment of organizational readiness, the development of internal resources, the design of a partnership that fits the organization and the ongoing management of the partnership. We have provided documented answers to the objections that are typically applied to ESI initiatives, and we have specified the conditions under which an ESI initiative will optimally perform. At this point it should be clear that ESI requires an approach that may well conflict with norms and traditions in many organizations. As a way of synthesizing the lessons in this chapter, allow us to draw a contrast between two approaches to supply relationships. At the heart of this contrast is the concept of trust, which will be explored in Chapter 7.

ESI cannot arise in an organization that is set on achieving benefits for itself at a supplier's expense. This is based on a fear that the supplier will benefit excessively and leads to a control mindset rather than a partnership attitude. Faced with such an attitude, suppliers typically react by adhering to the terms of a contract and making sure that nothing is provided that is not paid for. The end result is a 'win/lose' mentality: the manufacturer 'wins' only if the supplier 'loses' and vice versa. Conversely, ESI asks the partners to focus on the benefits available to both if trust and mutuality are invoked. This leads to a mindset of mutual support: both parties are encouraged to go beyond traditional thinking and prioritize the size of the total outcome ... as opposed to one's share of it. This is the meaning of the 'win/win' approach.

6 The Supplier's Perspective

The 'Push' Versus the 'Pull' of ESI

The FM Corporation is a foam molding supplier that posts the following message on its homepage (http://www.fmplastics.com):

> Recently a customer came to us with finished drawings for a quotation. Their design was unique and creative – but impossible to mould. Our engineers and toolmakers, together with the customer, eventually solved the problem. But earlier involvement with FM could have saved hundreds of man-hours and thousands of dollars.

FM is a supplier that promotes ESI with its manufacturing customers. It claims that ESI saves manufacturers time and money, and results in quality parts, low total programme costs and the shortest possible time-lines. Like a growing number of suppliers in its own and other industries, FM is using ESI to competitive advantage. Rather than being 'pulled' by innovative manufacturers into collaborative design work, it is 'pushing' the concept into its marketplace. This is in some contrast to earlier patterns of supplier–buyer transactions and, perhaps, signals a trend. FM's ESI programme includes early design meetings, site visits designed to establish personal working relationships, and FM-led partnerships with its own toolers, prototype companies and other manufacturing processors for secondary and subassembly work.

Seitz is a precision plastic moulder that concentrates on the electronics industry with customers mainly in the office automation business. Seitz is also organizing its operations around ESI and using this approach to promote the company. It claims that the ESI-based coordination of functions between a supplier and a manufacturer delivers products that meet specifications exactly:

> Through Early Supplier Involvement (ESI), Seitz works with your design engineers, to help design 'manufacturability' into every part. Once your design is complete, your CAD drawings can be transferred using IGES or DXF formats directly into Seitz' CAD/CAM system, through which we control the entire manufacturing process. Seitz engineers develop appropriate procedures and controls to assure product quality and uniformity.[105]

Plastech Corporation[106] has full-time ESI engineers that form a product development team with its customers. It claims these ESI-based teams develop better products, in less time, at a lower cost. Plastech promotes ESI with its clients and says the value added occurs in the following areas:

- Material Selection
- Tool Requirements
- Production Requirements
- Machine Process Capability
- Quality Standards
- Prototype Requirements
- Design Input
- Reduced Design
- Changes Manufacturability
- Finishing Alternatives
- Packaging Requirements
- Design for Assembly
- Cost Savings
- Model Requirements

FM, Seitz and Plastech are examples of suppliers on the leading edge of the changes that are redefining manufacturing systems. Where previously manufacturers encouraged suppliers – through argument and financing – to step into the design phase of new product development, these three suppliers have made ESI a core competence they push into the market. The timeliness of this development is brought into relief by the following passage:

> To make [ESI] … work, companies need to help their suppliers develop and maintain advanced technical capabilities. They must encourage their suppliers to invest in technology, but should be prepared as a last resort to take on new suppliers when their existing ones are not able to keep pace with developments.[107]

FM, Seitz and Plastech represent a new breed of suppliers that are already ahead of this advice. They represent a growing number of supply organizations that house the kind of specialized knowledge many clients either prefer to outsource given their strategic sourcing initiatives, or which they simply do not have in-house. This chapter explores the issues affecting this group of suppliers, and those others who wish to develop ESI capabilities.

Supply in the Strategic Sourcing Era

Manufacturers have followed several paths in their decades-old search for a more efficient way to manage the overall supply chain. Materials Requirements Planning (MRP), MRP II, Distribution Requirements Planning (DRP) and Enterprise Resource Planning (ERP) are recent examples of programmatic attempts to maximize the benefits flowing from an analytic model of manufacturing organization. Today, the importance of taking an enterprise-wide view of the elements and activities that go into a product is recognized by leading manufacturers. With this perspective, we have entered an era of 'strategic sourcing' that emphasizes the importance of maximizing the effectiveness of a **manufacturer's** resource flows.[108]

This chapter inverts the logic of strategic sourcing to outline the key aspects of ESI-based manufacturing relationships from a **supplier's** perspective. The manufacturer's view on strategic sourcing is expressed by Jon Ricker, Case Corporation's manager of purchasing operations: 'As organizations move further into strategic sourcing, they select suppliers that know their expectations and fulfill them on a regular basis.'[109] But as FM, Seitz and Plastech illustrate, one group of suppliers is not only anticipating the client's expectations, it is co-opting these expectations by introducing ESI-based interdependency as a key feature of the supply relationship. This is a significant departure from earlier arrangements.

The traditional Industrial Marketing/Purchasing framework regards suppliers as relatively passive parties in manufacturing relationships. The conceptual point of departure is usually a finished product that the manufacturer produces, for which parts and components are solicited from suppliers. The manufacturer controls design, specifications, delivery and to a large extent, pricing. Excellent suppliers are those that faithfully execute the manufacturer's design specifications and provide high quality parts or components according to schedule. The supplier is therefore placed in a secondary position to the manufacturer in terms of expertise, manufacturing discretion and bargaining power. This is a function of the manufacturer's position in the 'product development food chain', which allows more direct access to the end-user (or need-defining, hence product-driving) market. It is because of this position that most of the practice and academic literature has focused on the manufacturer's (buyer's) ability to structure, rationalize and develop the supply base.

But Industrial Marketing dynamics make it clear that suppliers are manufacturers in their own right and that the terms 'manufacturer' and 'supplier' have been rather arbitrarily defined relative to a traditional product development value chain, the anchor of which is the finished product. While this has historically been the case, emerging conditions in today's product development marketplace call these assumptions into question. Like manufacturers, suppliers are companies that have functional, structural and systemic dynamics; they plot strategies within a network of customers, suppliers and competitors; they strive to maximize the benefits of their activities.[110] Like manufacturers, suppliers are positioned within an evolving product marketplace, the context for a network of relationships that define their business. It is clear that an interplay exists between the environment in which suppliers find themselves, and their strategic and organizational responses. ESI is one of these responses, a significant change in buyer–seller relationships that has far-reaching implications for suppliers.

The Supplier Context

A growing segment of suppliers is beginning to assume many of the functions and responsibilities that were previously the domain of manufacturers. Another segment is operating in a traditional mode by providing commodity-level parts or reacting to a manufacturer's specifications. To characterize the differences, consider the work process favoured by the FM corporation noted above. It suggests the following process to its ESI clients:

1. Early design review meetings with FM engineers and toolmaker
2. Site visits to meet with key personnel and establish working relationships
3. Develop the project timetable
4. Identify the FM Project Administrator
5. Coordinate CAD/CAM systems for fast, convenient data transfer
6. Initiate an Advance Production Planning Forecast
7. FM partnerships with toolers, prototype companies and other manufacturing processors for secondaries and subassembly work
8. Additional design services needed from FM and/or tooler
9. Identify potential problems and make contingency plans

FM's outline advocates a high level of ESI involvement. The emphasis on supplier–manufacturer product development teams, electronic data transfer, partnerships and design-level involvement indicates ESI at its more advanced levels. ESI activities exist on a fluid continuum, however: not all suppliers operate at the same level of ESI involvement. It is possible to categorize ESI supplier activity in this regard according to the degree of design-level involvement that we defined in Chapter 2 (Figure 2.4) and which we further develop in Figure 6.1.

In Level 5 ESI, suppliers assume full responsibility for providing manufacturers with a complex system or subassembly that meets the manufacturer's needs. An example is Bosch and its ABS braking systems. Car makers rely on suppliers like Bosch for their ABS systems due to the complexity of the product, the expertise required to produce it in relation to the final product, and the development effort needed to stay current in these increasingly specialized fields. Suppliers operating at this level may hold patents that give them exclusive access to technology, insist that the designs they develop in the course of a project remain their intellectual property, and do not hold the manufacturer liable (in other words, the suppliers assume legal responsibility and consequences) for their products. These three features indicate the level of expertise involved and the importance, to the supplier, of guarding its intellectual property as a distinctive competence.

Suppliers at Level 4 of ESI have many of the same characteristics but produce more discrete parts or components that require less supplier integration relative to the finished product. The car manufacturers that expect Carello to design high quality headlamps that meet international regulations, for example, largely provide surface dimensions and contours

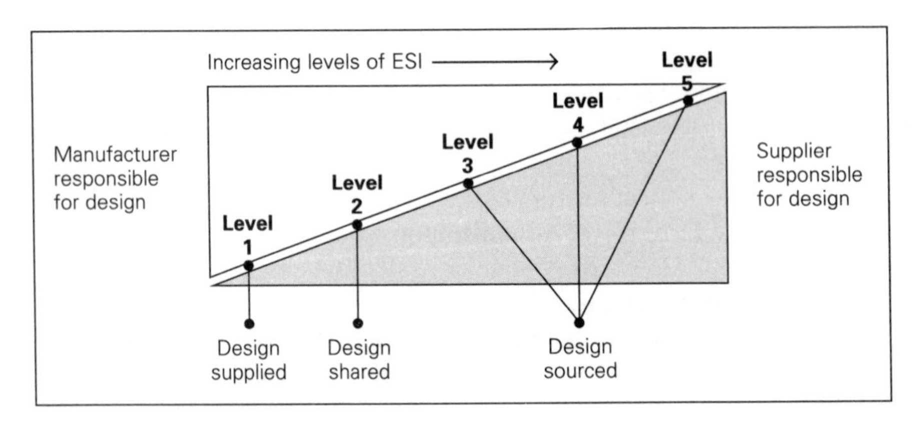

Fig. 6.1 Levels of ESI

around the headlight area as specifications. Ricardella Valente, Carello's commercial manager, told us that Carello first verifies a manufacturer's initial designs before prototyping a product and sometimes finds that the client has expectations that are impossible to meet. Development meetings are held in these cases, where modifications are worked out by the car maker in conjunction with Carello's lighting engineers. Carello then manufactures the lighting systems for which it holds the intellectual property rights and is held legally responsible.

Level 3 ESI is one where the supplier is only responsible for the development of a single part. In Level 2 ESI, however, suppliers and manufacturers are active product development collaborators. Manufacturers typically approach a supplier with rough specifications for a final product, and then engage in a collaborative/consultative process where the supplier's expertise is integrated within the overall manufacturing process. This is the level of ESI that locates the process outlined above by FM; it is also the one generally visualized when ESI is mentioned. The process is becoming classic: a manufacturer requires a product with exacting specifications, lacks the in-house resources (expertise, capacity) or has outsourcing objectives, approaches a supplier with whom a product development team is formed, and the development process proceeds to its conclusion.

Level 2 ESI suppliers might simply provide manufacturers with feedback on their specifications in a reactive fashion. This feedback is usually related to the cost or quality improvements a manufacturer can expect if certain modifications are followed. A typical scenario involves a manufacturer with significant in-house engineering capability that involves a supplier in discussions on existing designs, seeking improvement suggestions or specific information on how the part might be produced. At Level 1, the lowest level of ESI activity, suppliers provide manufacturers with feedback concerning their equipment, capabilities and production scheduling. The manufacturer is in control of the design at this level, as (s)he is at Level 2.

These five levels exist on a continuum that distinguishes ESI-oriented suppliers from those operating in more traditional ways. Many observers believe that there is a long-term trend to push design activities out to suppliers; at levels 3 through 5 of ESI, design responsibility is squarely on the supplier. Our research programme found that roughly 40 per cent of the suppliers interviewed operated in these higher or 'designed sourced' levels; the other 60 per cent operated at levels 1 or 2. The suppliers operating at levels 3 through 5 tend to deal with complex products and rapidly evolving advanced technologies. They concentrate their high levels of expertise which, owing to their internal organization, are readily accessed by outsourcing manufacturers to their ultimate advantage (higher quality, lower overall price, faster development).

Organizing for ESI

What are some of the characteristics that distinguish ESI suppliers? Our research and that of others indicate that three groups of organizational factors – pertaining to strategy, organization and operations – distinguish ESI suppliers from others.

Strategy

Suppliers that engage in high levels of ESI have generally positioned themselves in complex, rapidly evolving product market sectors. We noted in Chapter 4 that manufacturers are faced with increasing product complexity, technological change and manufacturing complexity. ESI suppliers tend to be located in business sectors where manufacturers have found it difficult to keep pace with product proliferation and the resulting operational complexities. This follows a clear economic logic: as products proliferate and become more complex, manufacturing systems reach a level of resource intensity that makes outsourcing attractive. ESI suppliers have therefore made a strategic choice to position their business activities in sectors where complex components and subassemblies place heavy demands on a client's resources. Some, like Seitz, refuse to work otherwise: 'We simply walk away from business that isn't high-end ESI.'[111]

Consistent with this product market positioning, suppliers working at the higher levels of ESI have generally made a strategic decision to focus their activities on a set of core competencies. No less than their manufacturing counterparts, suppliers operating in the current economic context feel pressured to concentrate their efforts on high value-added products and services. Companies like Carello, therefore, consider themselves the experts in their domain and work with manufacturers on this basis. 'Ten or fifteen years ago, Carello had much less authority with car manufacturers, receiving only the assembly programme for the lamp from certain Italian customers. In the past ten years, this has changed completely: the customer now understands that Carello is the specialist in lighting.'[112] Suppliers like Carello, then, have strategically focused on core competencies that respond to the outsourcing strategies of their clients. What this requires, as we will discuss shortly, is a continually developing expertise in outsourcing-intensive areas, such as microelectronics and plastics, and internal systems and structures that make this expertise accessible to manufacturers.

Focusing on core competencies involves trade-offs and choices; price, quality and speed are three variables that ESI suppliers juggle at all times. While cost is always a factor, suppliers in our research generally found that higher product quality and faster development cycles were the primary benefits sought by their manufacturing clients. This is logical given the pressures manufacturers face to shorten their development times while improving product quality and features. Nypro's key contribution to a new Gillette women's razor, for example, was quick debugging of the tooling which allowed Gillette to exceed its objective for product launch. Nypro regularly develops, debugs and launches new products in short periods of time; Gillette regularly manages long-running products but lacks engineering focused on product launch, and in this particular case, Gillette had a time frame of 12 to 18 months to launch the product. Nypro promised it would cut the time in half and beat the estimate by three or four months. This was the first time in 120 years that Gillette gave total moulding and assembly of a product to an outside company. 'Nypro's contribution on this project,' said Joseph Rizzo, Director of Programme Management, 'was speed.'[113]

Organization

Suppliers with ESI experience have developed systems and structures that support collaborative forms of working with their clients. Among the most important is a project-based organization that permits regular interaction with a manufacturer's representatives. The cross-company project team is a concrete embodiment of the ESI concept and takes different forms: from work distributed over geographic distance with no face-to-face interaction to 'in-plant' systems where supplier representatives are physically housed on the manufacturer's premises. The traditional purchasing function is itself subject to cross-functional organization: 'purchasing and supply management personnel in the future will work "outside" the purchasing department as part of integrated, cross-functional teams.'[114] Details change from supplier to supplier and project to project – including the project's leadership, its duration, its functional and employee composition, its hierarchical arrangements and so on – but the creation of a project entity to undertake the job is invariant.

A recurrent theme in this book is the need for openness and trust between suppliers and their clients. Many ESI suppliers report that they have an informal ethic of providing relatively free flows of business and technical information to their clients, and that they expect the same in return. This informality needs to be integral to a supplier's organizational norms and values, the 'soft factors' that comprise a system of expectations for professional conduct. A consultant in this area told us that secrecy is often a barrier in supply relationships but that the 'code of shutting up in the car industry as a supplier' is dissolving.[115] Contracts are insufficient to fully ensure confidentiality but openness and trust can be structured explicitly into the arrangement. FM and Nypro, for example, use site visits and other devices designed to personalize working relationships. Seitz makes its labour and materials costs available to a client and welcomes suggestions.

Industry observers believe that ESI suppliers are increasingly becoming 'systems integrators' that relate to secondary suppliers on one side of the value chain, and Original Equipment Manufacturer (OEM) clients on the other. This has been the case in the automotive industry for some time, first in Japan, then North America, and it now appears to be a growing trend in Europe. OEMs increasingly deal only with a first-tier systems supplier. Over the past five years, for example, seating suppliers in the automotive industry have reduced in number from ten or more to three first-tier suppliers that increasingly practise ESI vis-à-vis their own suppliers. In these firms, cross-company project organization and a relatively free flow of information within the system tend to characterize relationships in the value chain. ESI suppliers report partnership protocols with secondary suppliers such as Carello, which often designs tooling together with its secondary suppliers. The long-term trend, then, is to organize for ESI on both sides of a supplier's value chain: with company supply as well as with its demand.

Operations

Since an increasingly larger share of product costs are being outsourced, purchasing and supply are becoming key factors in manufacturing cost management. Firms are shifting from standard costing practices that emphasized an allocation of overhead costs, to a heavier reliance on activity-based costing, target costing and strategic cost management concepts.[116] Within this context, ESI suppliers are developing a variety of ways to deal with the economic side of contracts with their clients. An issue that should be raised here is that ESI opens the possibility of a supplier being the developer of a product and not its producer. A supplier may invest significant effort in developing the design for a component with a client which the client then takes to another supplier for production. Although rare, this is a valid concern and it is advisable to make expectations on remuneration clear at the outset regarding design, development and production. Sonceboz, for example, often has a 'gentlemen's agreement' with clients during the feasibility stages of development and formalizes the relationship as the project moves closer to production. This informal agreement may cover a variety of matters depending on Sonceboz's working relationship with the client, including an eventual specified target price. Other suppliers, like Nypro, prefer to specify design, development, prototype, tooling and production costs independently. While difficult to standardize due to the wide variety of situations and projects, some of the variables that may affect ESI supply contracts are outlined in Table 6.1.

Related to the issue of economic accords are those involving intellectual property and legal responsibility. In these cases there does not appear to be a great convergence of practices among ESI suppliers. Some expect the designs they develop for manufacturers and in conjunction with their products to belong to the supplier, while others do not. Carello, for example, flatly expects all of its lighting designs to remain their intellectual property: 'Our written contract lays all this out clearly. There are no questions in this area,' said Engineering Manager Ricardella Valente.[117] Carello also assumes all legal responsibilities relating to their lighting designs and provides guarantees to manufacturers. Sonceboz's

Table 6.1 Variables in traditional and ESI supply contracts

	Traditional supply relationships	**ESI supply relationships**
Formality	Generally fixed: high formality	Varies: low to high formality
Stages	▪ Tender ▪ Bid ▪ Production	▪ Feasibility ▪ Design ▪ Development/Prototyping ▪ Tooling ▪ Production
Contract	▪ Price and reduction schedule ▪ Time-based renewal	▪ Target costing ▪ Life of part/component/system
Profitability factors	▪ Savings resulting from efficiencies kept by the supplier ▪ Supplier's margins uncertain	▪ Savings passed on to manufacturer ▪ Supplier's margins acknowledged

contributions similarly belong to Sonceboz and the company is protected for three years in exclusive supplier arrangements: 'Because the specifications are so precise, the know-how has to belong to Sonceboz. Sonceboz would not get involved any other way.'[118] Nypro, on the other hand, commits to agreements which give the client all rights to the technology developed in their ESI relationship. In return, Nypro expects exclusive production rights for extended periods of time (up to five years). The question of intellectual property obviously has an important bearing on the type of contract a supplier will develop with a manufacturing client.

The last operational matter pertinent to ESI suppliers as a group is that of information availability and geographic proximity. These are both manifestations of the fundamental need to converge two companies' development efforts in as little time and space as possible. Having common CAD/CAM systems that allow simultaneous engineering and data transfer, in-plant engineers that work with a client's engineers on the premises, and facilities located close to a manufacturer's site all contribute toward this integration. ESI suppliers in our research found it helpful – though not critical – to minimize the 'business space' separating them from their clients. Bose Corporation, for example, had offices for the staff of several first-tier suppliers on its premises. Similarly, Honda recently built 30 offices in its engineering section to house supplier representatives that work with the company on new product development.[119]

A Summary of Key Issues for ESI-based Supply

If we digest the issues explored here and synthesize a set of operational recommendations for the managers of supply organizations, what emerges? The following is a list of propositions that result from our research and experience.

The practice of ESI will proliferate among producers and industries. The trend seems clear: more industries and more producers are using ESI for a variety of reasons. Suppliers can expect to encounter more clients who ask for ESI services, and more situations where their own ESI capabilities lead to competitive advantage.

Suppliers will increasingly 'push' ESI into their producer markets. This is an emerging trend but also a logical outgrowth of two factors: the increasing stock of knowledge housed in ESI suppliers, and the economic incentives for producers to outsource the design of parts and components.

Suppliers who proactively market their expertise to the producer market will, it would seem, put themselves at a competitive advantage.

Suppliers that choose to develop ESI capabilities need to evolve and maintain a knowledge organization. Nypro's Vice-President Brian Jones told us that bringing ESI expertise into his organization felt like 'going into the zone of rarified air'. Suppliers that expect to contribute to new product development rather than responding to its outcome specifications must ideally exceed the technical expertise of their clients. This is a significant step for some suppliers, and well outside the parameters of their business model: 'All of a sudden, instead of draftsmen we needed creators!' The supplier is required to develop a thorough understanding of the technical domain in question, matched by in-depth industry knowledge in the producer's context.

Suppliers that choose to move into ESI will need to rethink their business processes. The classic ESI business process design model is well represented by FM, which we recap below:

- Early design meetings between supplier and producer
- Site visits for key personnel to establish working relationships
- Agreement on contracts and schedules
- Establishing project organization
- IT, in-plants and other methods to minimize business space and time
- Supplier partnerships with secondary suppliers and subassemblers
- Continuous monitoring of plans, problems and schedules

The supply organization's basic move is from a traditional and reactive posture to one of active involvement and continuous implication in the project. This has organizational/organizing implications. A clear implication is the need for project leadership, or a champion. This champion will normally be a member of the producer's organization but not always, and when a vacuum exists on this count the supplier should be ready to take corrective action. Another organizing implication is the need for quick and efficient communication between organizations. IT is currently providing major advances in this area but suppliers who are not so equipped, or expert, will need to come up to speed quickly.

Suppliers engaging in ESI will, as a result of contracting, face less economic uncertainty than others. This proposition is made in a complicated and under-researched context: ESI-based supply contracts versus traditional supply contracts. But the existing evidence indicates that because cooperation rather than competition is the basis for negotiation between a supplier and a producer, the successful ESI supplier is assured of lower cost vulnerability and profit variability. Likewise, the chance for disproportionate profits is removed given the openness with which financial matters are discussed: disproportionate profits, in the form of cost savings, are expected by and passed on to the manufacturer.

Suppliers developing new products or innovations in an ESI context will increasingly be faced with intellectual property issues. That this is a logical outcome of a supplier's creative genius in a project does not diminish the difficulty with which intellectual property claims are discussed and managed in business. Genuine innovations provide their owner with profound business advantages in the market for a period of years. Suppliers that implicate themselves in the development of innovations should have a clear strategy concerning intellectual property.

7 Managing Interdependence

In 1992 Philips Floor Care, one of the leaders in the European vacuum cleaner market, realized that ESI was a must if they wanted to accelerate product development at a time when their business was confronted with a catch-up game.[120]

A new type of machine, the 'injection cleaner', was invading Philips' market. An injection cleaner combines a wet-and-dry vacuum cleaner with a shampooing function, and is also called a 3-in-1. This product had been introduced to the UK in 1987 by an outsider to the industry, a company called VAX. Initially all leading vacuum cleaner makers, Philips included, displayed scepticism about the chances of this concept, though they were not taken by surprise: VAX had initially tried to license the technology and then went into business themselves when they failed to find a partner.

After a few years of watching the injection cleaner sales grow, particularly in the UK, Belgium and France, Philips Floor Care and other major vacuum cleaner makers decided to enter the market. Philips agreed that in the beginning, for the sake of rapidity, entry would be through an arrangement with a small OEM supplier who had already developed a 3-in-1. At the same time, capitalizing on the market and technology intelligence gathered by the OEM operation, Philips Floor Care decided that it would design a totally new machine in order to leapfrog existing competition.

The launch took place amid the Centurion operation, a corporation-wide initiative that aimed to re-energize Philips after the substantial restructuring led by Jan Timmer in the early 1990s. Wim Van Der Berg, the head of the Domestic Appliances business group, and Carel Taminiau, general manager of the Floor Care group, agreed that the injection cleaner project was a good opportunity to mobilize the organization and put it back into a competing mood. A target date was formally set: market introduction to the trade in June 1993, comprising a total development cycle (from concept to delivery) of 18 months, compared to the traditional 30 to 36 months. To emphasize the link with the Centurion Operation, the product was given the Olympian name 'Triathlon'.

A product development team was put together, under the joint supervision of a young and dynamic product manager, Henk De Jong, and a seasoned and highly competent engineering manager, Hugo Van Der Woord. These two individuals were quite complementary, one providing the team with the high level of energy critical to ensuring a sense of urgency, while the other guaranteed rigour in the development process.

Both project leaders realized that they would need to break away from conventional approaches if they were to meet the delivery date. Looking at the traditional process, it was obvious that transactions with suppliers were creating important delays. Several weeks could be gained immediately if suppliers could work in parallel with the Philips engineering team. Suppliers were needed to manufacture several plastic components of the injection cleaner. Thus, the decision was made to invite suppliers as early as possible to discuss design issues.

This decision represented a substantial departure with the past, because Philips Floor Care had traditionally put pressure on its suppliers, playing them against one another through competitive bids. The resulting atmosphere was somewhat adversarial and had not generated a great deal of mutual goodwill. Thus, when the team had to select a supplier to work on the design of the new product housing, they were reluctant to call on the usual subcontractors, fearing that they might not play the game fairly. Philips would have to share confidential information, thus exposing itself to potentially opportunistic behaviour on the part of the supplier.

What was missing is obvious: trust. Philips Floor Care did not have enough trust in its plastic parts suppliers. Philips managers intuitively felt that trust is an essential ingredient of a successful ESI relationship. They realized that, because of the suppliers' involvement in the design, they would become dependent on these firms (as we explained at the end of the first chapter; see Figure 7.1 on page 156), and did not think that they could accept this considering the type of relationship they had typically had with their supplier base. Consequently, Philips decided to approach a new supplier, one with whom there were no uncomfortable memories, and to start a fresh relationship based on openness, reciprocity ... and hopefully trust.

This chapter, building on the preceding conclusions, will provide a reflection on the nature and origin of trust. Our aim is to achieve more than a discussion of this fascinating, yet little understood, concept. Based on both academic research and our own observation, we will offer suggestions on how managers should approach ESI so that the chances of a trusting relationship are maximized, and interdependency is managed as effectively as possible.

We will first explore why the importance of trust is perhaps underestimated, through a discussion of dominant views in economic theory. We will show that this attitude can be attributed to a perspective which looks at the transactions, but ignores the specific partners involved. We will then illustrate the positive role that trust can play in buyer–supplier relationships. Two sections will be devoted to a review of recent academic developments in this area. These will help us extrapolate a model for managing trust in ESI relationships.

Is Trust the Rule or the Exception?

The idea that trust is essential to business relationships is quite controversial. It has strong supporters and strong opponents. The supporters assert that trust is the very essence of business life and that no economic activity could be conducted in its absence. The detractors mock this view as naive, claiming that trust is more the exception than the rule because humans always tend to take advantage of situations when there is gain to be had. They believe that rather than developing trust, economic actors should seek protection through rigorous contracts.

As a matter of fact, the dominant view in economic theory is quite consistent with the sceptics' view. Transaction Cost Economics, which has become a dominant model for the analysis of relationships between firms, is based on the presumption that economic actors are opportunistic, that they lack candour and honesty in their dealings with others, and that consequently, in certain circumstances (high level of uncertainty, asymmetry of information, limited number of competing options, and so on) they would quite probably try to take advantage of the other party. This model thus suggests that economic actors are primarily seeking to reduce the costs of this opportunistic behaviour through the development of appropriate governance structures, such as hierarchical organizations and contracts. If a company, for instance, decides to internalize a transaction (that is, to make rather than to buy a given item), this is because it is both too risky to trust external agents and too costly to try to prevent the opportunism of the other party through complex contracts.

The promoters of Transaction Cost Economics (TCE), and Oliver Williamson in particular, have not considered the other option offered to contractors: to conduct a given transaction with someone they trust. Actually, when confronted with this idea, the supporters of TCE argue that one has to assume the presence of opportunism to understand how governance structures are formed and how they operate.[121] Opportunism is seen as a basic condition that can explain the dynamics of institutions, just as gravity is a law of nature that explains how planets move in relation to one another. Any departures from this basic condition are to be understood as an exception, an insignificant noise in the system.

The TCE model is not interested in specific conditions where actors try to reap economic advantage by finding ways to prevent, or minimize, opportunism. Yet, some firms are actually built on this very principle, operating through a network of partners, and they surpass their competition because they do not have to bear, like other competitors, high transaction or organizational costs.

TCE is not suited to explain this type of situation because it focuses on a set of general objective economic conditions surrounding a transaction (the level of uncertainty, the number of possible partners, the distribution of information between the contractors, and so on) but not on the identity of the parties involved. Trust is, by definition, not an objective condition since it subjectively bonds specific actors. In a potentially risky endeavour, those who have a trustworthy partner have a far better chance of succeeding than those who do not. Trust is thus actor-specific. But TCE does not consider the subjective conditions of a given transaction: it ignores partners' identities and trust is therefore subsumed in favour of objective economic 'realities'. This is because TCE's purpose is to analyse and define what type of governance structure is appropriate given certain conditions, regardless of the parties involved.

The reality of trust is quite different. The same actor could behave opportunistically with one party, and fairly with another, even within the same transaction. We know intuitively that pre-existing social bonds are critical to determining the attitude in a transaction. So, not only is trust a possible option, but opportunism is not as general as assumed by transaction cost economists since it can vary for the same individual in similar objective conditions.

The social context is ignored by TCE and, for the same reasons, the model ignores history. It does not take into account the dynamics of a relationship between two parties over time. On the contrary, TCE is essentially a static view of transactions, measuring the level of opportunism at a given moment, while trust is a dynamic phenomenon created over time through interactions between economic actors.

We need to go beyond objective conditions in order to understand fully how actors structure their relationships. This is certainly necessary when one wants to make prescriptive recommendations as opposed to broad observations. In this light, trust should not be considered a peripheral phenomenon that has no bearing on the analysis of ESI. We will argue that it always matters, even when (perhaps especially when) it may be lacking. Although not universal, it is possible that, given its benefits, partners in a transaction are likely to pursue a modicum of mutual confidence. We will see in the next section that trust is indeed critical in ESI relationships.

The Role of Trust in Buyer–Supplier Relationships

Trust has recently become a fashionable topic of discussion with popular writers like the controversial Francis Fukuyama[122] and Alain Peyrefitte,[123] who have been debating the differences between countries in terms of the propensities of various people to extend their trust to others.

Both writers argue that the economic development of a country depends on the ability of its people to trust partners in commercial transactions. According to these authors, there are low-trust and high-trust countries. For Fukuyama, a country's trust level depends on the radius of trust, that is, the area of the community within which people are able to give their trust. In some countries this area is limited to family ties (China, Italy, France) and therefore enterprises tend to be of limited sizes. In others (Japan, Germany, US), the radius of trust is much broader, allowing for the emergence of much larger industrial organizations. Peyrefitte comes to similar conclusions when he compares the Protestant background of Northern European countries, which is supposed to encourage self-confidence and trust, to the Catholic cultures of France and Italy which promote a distrust of money, entrepreneurship and innovation.

Thus, these two books closely associate economic success and ability to trust. They also warn us that our societies increasingly tend to promote distrust, hence re-creating the conditions which have delayed industrial development in certain countries and possibly hindered their ability to compete in the world economy. Developed countries are encouraged to return to culture and values that allow the emergence of trusting relationships.

The propensity of people to trust others is certainly influenced by national cultures. We will not debate this. But undoubtedly, within a given country, there are substantial differences between individuals in how they initiate and maintain trust. Numerous factors are in at play at this level, and these are clearly relevant to an understanding of the governance structures adopted to manage economic transactions.

All economic activity requires the coordination of actors who participate in a production cycle. For instance, in a vertical chain of activity, the producers who operate at the different stages of the chain need to adjust the quantity they make so that the upstream activities supply enough, but not too much, to the downstream ones. How can this be ensured? Bradach and Eccles note that there are three ways to perform coordination: price, authority and trust.[124] Trust is thus another mechanism for achieving the control of business activities. For a given transaction, coordination can be achieved with two contractors simply agreeing to perform the activity in question.

When it comes to complex relationships such as ESI, trust is particularly relevant. Authority cannot be the primary mechanism precisely because the client has decided not to perform the design internally and prefers to involve an outsider. Price is also not adequate because, as explained previously, there is generally no clear yardstick to help the buyer assess the value of the design service provided by the supplier. At best, there are other options, submitted by other suppliers, but typically the proposals are quite difficult to compare.

Moreover, the conditions that prevail in ESI relationships are such that a sufficient level of trust between partners is needed above and beyond the contracting possibilities. These situations would make detailed contracting very expensive, and perhaps not even possible, as several ESI partners experienced. Some of those situations are outlined below.

- **High uncertainty about outcome** The goal of ESI cooperation is often quite difficult to express in simple contractual terms, and certainly more complex than for a traditional supply arrangement. The client expects the supplier to provide design ideas that will improve its own product, but cannot define what is expected because they have handed over the responsibility of defining those improvements to the supplier.
- **Asset specificity** ESI typically calls for the development of assets specific to the relationship. The supplier, for instance, has to invest a substantial amount of time learning about the client organization. This is an asset that would be lost should the relationship be discontinued. The same can be said about engineering work that the supplier contributes: it only has value as long as the project goes on.
- **High information asymmetry** The relationship actually arises from an asymmetry of expertise between the client and the supplier. If the client had as much information as the supplier there would be no need for an ESI agreement.
- **Medium- to long-term relationships** The ESI partners will have to depend on one another for a period of several months during the design

phase and possibly several years when the product goes into production. Throughout the length of the project, they will have virtually no alternatives or, at best, limited ones.

According to the supporters of Transaction Cost Economics, these conditions should generate a high level of opportunism, possibly making the agreement so expensive to control that the client would look for possibilities to internalize the relationships, hence negating the benefits of Early Supplier Involvement.

This point is particularly relevant because the contributions of each partner in the ESI relationship are typically very difficult to assess. Sometimes, the benefits can be estimated afterwards, in terms of the profit derived from the joint innovation – for instance, when the supplier brings in a technical solution that can be compared with a previous one. But this is not always the case, and as we know from research on innovation, most of the time it is difficult to identify, much less evaluate, the factors that explain the success of a new product. Consequently, the partners might never really know how much each contributed, and thus it is likely that the profit sharing will be decided on subjective impressions as opposed to objective documentation. Thus, perception and subjectivity typically dominate these types of relationships, making them even more susceptible to opportunism.

But internalization is not the only alternative. There is another: that trust is sufficiently high between the partners. If the two partners have trust in one another, they will assume that the other partner will not behave opportunistically, and consequently will accept conducting the transaction as independent parties. In other words, given the above-mentioned conditions, ESI relationships would not succeed unless trust pre-exists. At this point, we should define the concept of trust.

The Nature of Trust

Trust is often defined in ethical terms. Most economists such as Arrow[125] and Williamson,[126] as well as management specialists like Bradach and Eccles[127] and Ring and Van de Ven,[128] would define trust as the presumption that another party will not be opportunistic or dishonest, and therefore will not perform an action that would be harmful to others.

There is no question that trust has an ethical dimension. Trusting a partner means, in particular, believing that s(he) will abide by general ethical standards. There are numerous ESI situations in which the need for ethics can be identified. Suppliers to GM, for instance, once complained that their client had shown engineering drawings to other suppliers in an effort to get better contractual conditions.[129] The dispute between Braun and their glue supplier over the coffee pot handle innovation is of the same nature: one of the partners considered that the other did not behave ethically.

But honesty alone is not enough to be considered trustworthy. What good is a partner who is honest but does not have the required know-how? As a matter of fact, contrary to what one might expect, opportunism is not the primary source of apprehension in buyer–supplier relationships. When asked about their greatest fears in the context of ESI relationships, the manufacturers we surveyed did not mention a lack of honesty on the part of the supplier as a major problem. Much more important for them was whether the supplier had the necessary capability to deliver what they, as clients, needed.

As explained by Bidault and Jarillo in a recent book on trust in economics and management, trust has a technical dimension.[130] Having trust in a surgeon normally does not mean expecting that (s)he is honest. It typically means believing that the surgeon will do the job as well as possible, given the state-of-the-art of surgical techniques and the physical condition of the patient. In other words, trust concerns the expectations that the partner has the required competence to deliver what (s)he has promised.

In the ESI context, we find countless examples of this type of trust. For instance, when Fuji Xerox decided to contract Sango Kogyo to manufacture the metallic frame for its new high-speed colour copier (DocuColor), they had sufficient trust in the technical competence of this supplier. Fuji Xerox actually spent a considerable amount of time documenting this competence through in-depth assessments of the technological capabilities of the possible suppliers. The selected partner was supposed to reach an exceptionally high level of precision because the colour copier design had very tight tolerances. In fact, Sango Kogyo at the time did not fully meet the expected performance level, but Fuji Xerox executives trusted that, with adequate support from their engineering department, the supplier would eventually be able to build the necessary competences.

This example also indicates that trust is more than a belief in the technical knowledge of the partner. There is also an expectation regarding its attitude. Fuji Xerox expected that Sango Kogyo would be willing to improve its competence. Competence in this type of transaction clearly goes beyond technical expertise, and includes the ability to interact with others. This is what Mari Sako calls 'goodwill trust'.[131]

The relationship between Lexmark and Molex on the Liberty project almost collapsed for this reason. Molex, the supplier of electrical connections, was supposed to steer the development with one its engineers, Jeff Tillou, who supervised Lexmark's technical staff assigned to this task. Initially, Jeff did not fully realize the extent of his role and did not take the lead, letting the Lexmark staff loose. Obviously neither Molex nor Jeff Tillou were lacking the technical competence. They just had not yet adopted the attitude expected by Lexmark. It took a serious discussion between Greg Survant, the project leader, and Jeff Tillou and his management to set the relationship in order. Only then could Lexmark trust that Molex would be able to deliver what they had agreed.

This view of trust is quite consistent with Charles Handy's recent article in which he described trust as confidence in someone's competence and his or her commitment to a goal.[132]

We submit that trust is a multidimensional perception that goes beyond ethical standards. We agree with the definition proposed by Bidault and Jarillo: trust is the presumption that, in a situation of uncertainty, the other party, even in unforeseen circumstances, will act in accordance with the rules of behaviour that are deemed acceptable.[133] Several implications arise from this definition, as follows.

First of all, trust is a presumption, or a belief. It is thus subjective, the perception of an individual. Consequently, one individual might trust someone that nobody else trusts. But for trust to be effective it has to be mutual. Therefore, the nature of trust is specific to a set of partners. A given transaction might be objectively quite rewarding, yet risky from an ethical as well as a technical standpoint, and very few parties might have enough confidence in each other to undertake it. Those who do gain an advantage over the ones who cannot find a trustworthy party.

A second implication is that the ability to trust a given individual differs greatly because it depends on the frame of reference of the person concerned. Ethical norms differ by community, tradition and context. Technical competence certainly varies with time and also, to a certain extent, place. Finally, the ability to trust will be shaped by personal preferences. Consequently, where one person might feel mistrust, another would have the opposite feeling because they refer to different standards. Perhaps this is why trust is often easier to reach between individuals that belong to the same community.

This also implies that trust can be extended to a partner for one transaction, but not for another one if the partner lacked the required competence in that situation, even though (s)he would be equally ethical. So trust is transaction-specific, in addition to being partner-specific.

Our definition also suggests that trust is a matter of degree. Because it is multidimensional, an individual might be trusted on one dimension but not on others. Furthermore, trust might be limited, with one party retaining some minor fears about the outcome of the transaction. In this case, (s)he can use a formal contract as a safeguard for situations of uncertainty. Trust should not be seen as a replacement for contracts, but rather as a complement, an emergency brake of sorts.

However, in situations where trust is low between partners, they will need to specify their expectations in great detail and to monitor, perhaps with the help of a third party, the actual delivery of these terms. The lack of trust thus exacts a cost which might be substantial. But, as Arrow observed, trust does not have a price since if one could buy trust, the mere desire to buy trust would immediately raise doubts about the trustworthiness of the other party.[134]

Finally, trust does not necessarily imply empathy or altruism, contrary to what some writers say. We might actually have confidence in people whom we do not like as long as we believe those people behave correctly and will not take actions that could compromise our interests. Similarly, trust does not automatically imply commonality of interest. It is obvious in any buyer–supplier relationship, and ESI arrangements in particular, that the two parties have opposing interest: what is revenue for one is a cost for the other. Yet, there might be mutual confidence if both believe that the other party will not try to pursue their own interests to the exclusion of mutual gain. Consequently, trust does not exclude opposition and possibly some type of conflict. What matters most is how conflict is resolved, that is, whether it is negotiated with the interests of each party in mind. Actually, it is in situations of conflict that trust is best put to the test. One aspect of trust is presuming that, should a dispute arise, the other party will behave fairly, and will do its best to preserve each side's interests.

We may now conclude that a trusting relationship is a mutually positive perception regarding the various capabilities of both parties to execute a transaction according to each sides' expectations. It does not exclude the existence of a contract for exceptional circumstances, but typically allows the scope of such formalities to be limited, and thus also the cost of contracting. Nor does it exclude the possibility of conflicts which are expected to be solved fairly, that is, consistently with partners' interests.

Having defined the nature of trust, we can now turn to the complex question of its sources.

The Sources of Trust

As explained by Bidault and Jarillo, many methods are suggested in the literature on how to bring about trust.[135] As we review them, we will notice that there is some confusion between methods that are trust-generating and those that are primarily designed to prevent the consequences of mistrust. In this section, we aim to develop an understanding of the origins of trust, which we will then apply to ESI relationships.

Ethics seem to be at the basis of trust because they guide individuals in their decisions, providing a set of standards for acceptable behaviour. Trust would thus be easier to develop between contractors if both parties believe the other side will abide by ethical standards. This perspective suggests that the existence of a strong mutual ethic, that is, one that everyone respects, helps establish trust between individuals. One implication of this approach is that since ethics are typically specific to a community, trust should be easier to build between individuals who belong to the same community or group than to different ones. In the current era of transnational business, it thus follows that trust might be more difficult to manage.

It is helpful to look into the role ethics actually play in the development of trust. Their major role seems to be as a prevention of opportunism. High ethical standards should reduce the probability of bad behaviour as a breach of those standards would have negative consequences. In other words, one assumes trust of another because the consequences of unethical conduct would be harmful to their reputation and thus for engaging in future transactions. Economists describe this as an 'exchange of hostages'. In a transaction, both parties entrust the other with an asset (their reputation) that would lose value if they did not deliver according to their promises. The most common view of ethics presumes they are effective on the basis of this 'consequentiality', or the fear of reprisals.

The same reasoning is applied by economists to contracting. Contracts could lead to a form of trust inasmuch as they bind parties to agreed conditions. If a partner does not adhere to these conditions, legal penalties would be applied. Therefore, goes the argument, if my partner accepts signing a contract, it means (s)he has good intentions. This argument is similar to the discussion of ethics. Parties would be exchanging hostages, albeit more explicit ones. Consequently, the same objection can be presented: the function of contracts is not so much to inspire trust, as it is to deter opportunism and dishonesty. Contracts serve to reduce the probability of misbehaviour, more than to generate good intentions.

In any case the 'good intentions' argument is only relevant in the case of ethical trust. When it comes to technical trust, it does not work. Subcontractors might well sign a contract in good faith for a transaction that goes beyond their competence. If this is the case, which is more common than disputes on ethical issues, contracts can be viewed more accurately as a contingency plan if trust should be misplaced. Contracts typically cover exceptional circumstances in which the consequences of a miscalculation would be difficult to manage because of the level of expenses incurred. These are situations in which trust would not be sufficient: unlimited trust is, in practice, unrealistic.[136] Contracts, then, are not sources of trust but rather set the conditions for trusting relationships to be implemented in the event that trust disputes arise.

Another dimension of ethics relevant to a discussion of trust is self-interest. We can trust someone when it is in their best interest to uphold their end of the agreement; for example, when they would incur higher costs, both financially and socially. This is similar to the arguments about ethics and contracts: we trust people because they would be better off delivering on their promises. Partners will be reliable because they fear the consequences (contractual or social penalties) or because they anticipate future transactions. This argument is more consistent than the ethical one with the concept of technical trust. A contractor might hesitate to enter into a deal that might endanger his or her technical reputation with the other party, as well as in the market at large. This concept of reputation is necessarily relative to a set of accepted technical standards. If no standards yet exist for a given transaction, the argument does not hold because the partners might not see the risk entailed and/or might have trouble assessing the outcomes.

A fourth factor is often mentioned as a source of trust: experience. The repetition of transactions between partners could be a basis of trust, the assumption being that a partner who behaved properly in the past is likely to be reliable. Game theory demonstrates that trust can be developed between two players if they play an unlimited number of times. But experience is more concerned with the process of trust-building than how it applies to its origin. Experience develops knowledge, yet not always cleanly, as it might just be based on inference. Therefore, experience is perhaps more a way to accumulate and verify information on the diverse factors that form the basis of trust.

Recently, the concept of familiarity has been introduced.[137] The importance of family businesses as well as local networks provides intuitive evidence that closeness among individuals does seem to generate trust. Familiarity provides an intimate knowledge of individuals, in terms of their ethical standards, but also sometimes in terms of their competence and their attitudes. This is the only factor that is relevant to all three dimensions of trust that we identified. Being familiar with someone means knowing a person in detail, and subsequently being able to anticipate his or her behaviour, and to assess the fit between a given set of skills and the tasks at hand. Familiarity thus allows one actually to develop a model of another person's pattern of behaviour in a variety of situations. From such a model, a belief might then be formed regarding the ability of the partner to take appropriate actions in projected circumstances.

This discussion suggests that experience and knowledge of the other party is indeed a basic component of trust, from which expectations about future behaviour can be elaborated. The various factors that we have reviewed all may contribute in some way to the development of this knowledge. But it is important to distinguish the process of trust-building from the source itself. The following section will examine how trust works in ESI relationships.

Managing Trust in ESI Relationships

Our analysis is consistent with that of Lynn Zucker, who suggested that trust actually emerges from three different bases: (1) the process through which the relationship was built, (2) the characteristics of the partners involved, including personal background and ethnicity, and (3) societal structures such as professional certification whereby a trustworthy institution guarantees the quality of an agent.[138] This view of trust-building allows a broader scope that goes beyond the ethical dimension.

Zucker's model also offers a perspective on the process by which trust is built, which is both proactive and analytical. This analysis, we believe, applies well to ESI relationships. ESI candidates have to assess their potential partners' trustworthiness, and at the same time, they have to demonstrate their own trustworthiness, since only mutually trusting relationships are effective. There is no priority assigned to either of these two tasks since they must be conducted simultaneously (Figure 7.1).

Demonstrate One's Own Trustworthiness

Trust is a multidimensional concept, and trustworthiness must be demonstrated on all three levels that comprise trust in partnering: ethics, competence and attitude. In the context of ESI relationships, based on our

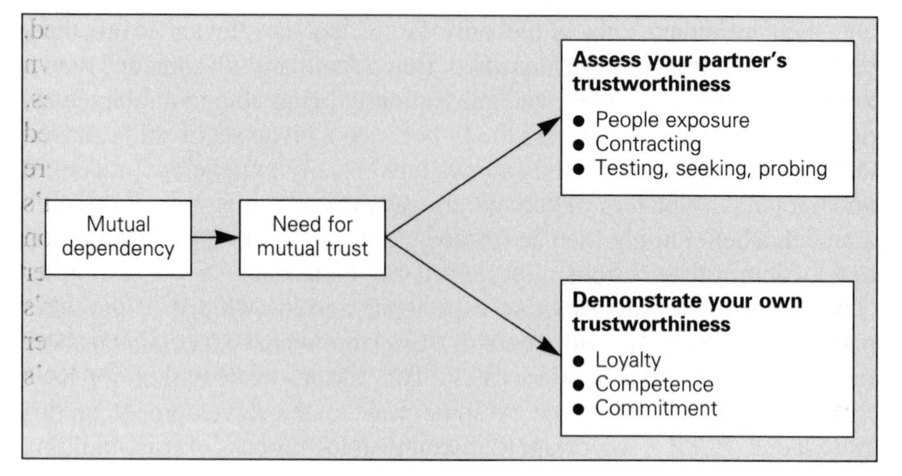

Fig. 7.1 Trust-building in ESI relationships

research, we believe that there are three distinct ingredients that demonstrate trustworthiness: (1) showing one's loyalty to the partner, (2) proving one's technical competence, and (3) providing evidence of one's commitment to the relationship.

To begin with, we should keep in mind that the trust-building process in ESI relations is not something that can be started from scratch. In most cases, the partners would have dealt with one another in the past, or at least they would know each other by their reputation in the marketplace. Therefore, it would be misleading to consider a demonstration of trustworthiness as a linear process that is launched at the outset of a relationship. Trust is typically evidenced through a succession of transactions, with the partner in question or with other firms.

Loyalty

What the supplier will certainly expect from the manufacturer in an ESI relationship is loyalty, that is, the expectation that the client will show respect for the supplier's agenda and the importance of its role. Intellectual property questions in particular raise the issue of loyalty and respect: the supplier will look for signs in the behaviour of the client that will provide evidence of that company's attitude to those questions.

From the beginning of the relationship, it will be critical to acknowledge the design contribution of the supplier, to provide assurances that the client does not intend to take over the ideas of his/her partner. At Renault and Matra Automobile's early meetings to discuss a possible cooperation on the manufacturing of Europe's first minivan, the concept was not finalized. Matra Automobile showed what they had developed on their own initiative. It looked interesting to some, but not all, Renault executives, and much remained to be done. Renault and Matra engineers worked together to create what would be known as the Espace. They were successful in making the best use of both parent companies: Renault's marketing flair and Matra Automobile's creativity. Although the decision was made to commercialize the car under the brand name Renault rather than Matra, Renault's management acknowledged openly their partner's contribution (although not in the car marketing campaign) and later allowed Matra to put a plaque inside the van to mark its role in Espace's development.

Competence

Because trust is very much about competence, significant attention should be given to demonstrating one's competence. Certainly the supplier has to show its competence, since its contribution will be essentially technical. But the client also has to show its competence. The supplier will want to know whether the manufacturer is likely to be commercially successful with the new product. After all, the supplier will need to invest a substantial amount of time and money and it is legitimate to expect a fair return on those efforts. Success, obviously, is never guaranteed. The supplier should, however, try to assess the capability of the client to achieve its objective.

When Salomon started the development of its Monocoque ski which required the use of pre-impregnated composite material, it turned to a French supplier for engineering support. Initially, the supplier was lukewarm, as it did not believe Salomon could achieve this breakthrough. Fortunately, based on its previous successes, the commercial reputation of Salomon was very strong. Nevertheless, it took quite a bit of demonstration on the part of the project leader, Roger Pascal, to convince the supplier that Salomon had done its homework and had studied the technology and the market in sufficient detail to make this project a success.

Commitment

Interestingly enough, the word for trust (shin) in Japanese is made up of two signs: hito which means man and gen which can either mean spoken or word.[139] Trust in Japan is therefore the equivalent of a commitment based on a person's word.

Because ESI relationships are at least medium term, if not long term, it is important for both sides to know that their partners are committed to the success of the project. A loyal and competent supplier, or client, is not enough if the project is not given adequate resources and appropriate priority.

Minco, the supplier to Lexmark, found itself in a situation where it had to demonstrate its commitment to a project when the client expressed its deep dissatisfaction with the development of the printer frame. A major misunderstanding had developed about the role that Minco was supposed to play in the management of the frame design and, as a consequence, the tool-making was significantly behind schedule. Minco reacted promptly, refusing to waste time on finger pointing, and John Levering, their head of engineering, took over the project and gave it top priority. This did much to express Minco's commitment to its ESI partners.

The analogy between alliance formation and courtship can be quite inspiring and has actually been used quite effectively in the literature on the strategic management of alliances (see, for example, Rosabeth Moss-Kanter's famous article).[140] To pursue this metaphor we believe that, just as in the bonding of a romantic couple, the first steps between ESI partners are very critical. In the beginning of a courtship the parties give away a lot of clues about their normal behaviour; these signals are read by the other, are progressively integrated into his/her mental model of the person in front of them, and become a means to assess whether this person is worth the engagement. At this stage, details count as they may be signs of more deep-seated tendencies. Therefore, lovers are careful about what they show about themselves. Companies entering a relationship as integrative as ESI need to do the same: be careful about the clues they give away and be attentive about the ones they observe. The next section will focus on these observations.

Explore the Partner's Trustworthiness

At the outset of an ESI relationship, or preferably before it starts, partners need to pay attention to what they observe in the other. This is not to say that trust is gradually built from the ground up, as if it were created out of nothing. On the contrary, we agree with what Morton Deutsch wrote in his classic 1962 article: that economic agents enter a relationship with the assumption that their partner is trustworthy, until proven wrong by observed behaviour.[141] This implies, as Buckley and Casson noted, that trust is as much the result of a relationship as it is a basis for it.[142] Therefore, trust is created through a succession of positive and negative interactions where its foundations are tested continuously.

Because the main source of trust is knowledge of the partner, exploring the trustworthiness of that partner calls for learning. The following three initiatives contribute to this.

Organizational Exposure

There is no substitute for direct observation in the exploration of trustworthiness. This is not simply because having your people meet the partner's staff will allow them to gather information about their habits, their competence, their ethics, and so on. It is also because mutual exposure contributes to the creation of trust. Personal interaction is a positive element in the development of trust among individuals because it allows an exchange of information regarding expectations and constraints.

Jeffrey Dyer showed that face-to-face meetings between Japanese manufacturers and their suppliers were substantially more frequent than at their American counterparts.[143] The number of hours spent by US car makers to meet face-to-face with their suppliers during a new model cycle was in the 800–1200 range, Nissan spent on average 3300 hours and Toyota over 7000 hours. The time invested in the relationship does result in better products, in terms of defects per 100 vehicles. Dyer reports the comment of a purchasing manager from Nissan: 'direct communication and relationships developed over a long period of time made detailed and explicit written communication largely unnecessary'.[144] We would argue that such a level of interaction also contributes to trust as it reinforces the feeling that supplier and client are actually a team pursuing a project jointly. As noted previously, familiarity is an essential ingredient to trust, and from this perspective face-to-face contact is probably a very effective method.

Information Sharing

The information collected by the people interfacing with the partner should be circulated to check for consistency and to help form a clearer image of what that partner is all about.

One of the major differences between Western and Japanese firms we have noticed is the thoroughness of partner selection. Japanese firms are known to maintain a base of suppliers with whom they work regularly. Their selection process might even look cumbersome to some Western executives. Fuji Xerox, for instance, has developed very detailed assessment forms which list more than 20 criteria with which they rate potential suppliers. The focus of the selection, by the way, is competence and attitude, with ethics very much in the background. This process is only applied to a few (three to five) supplier options. The benefit of this systematic approach is to have detailed information on the reliability and trustworthiness of the various suppliers to circulate among the managers involved in the ESI relationship.

Contracting

As we explained earlier in this chapter, trust should not be seen as a substitute for contracting. At best, it serves to simplify it, but ESI partners should not dispense with all contracts. Nor should contracting attempt to replace trust, because the agreement could become quite heavy with detail and as a result hamper the needed flexibility (see Chapter 8).

There are virtually no ESI companies which have not at least signed a confidentiality agreement. In fact, this is often one of those details that signal good intentions to the partner. This alone, however, is not sufficient. ESI partners should develop a short contract specifying the conditions governing the exploitation of the technology developed, as well as the remuneration of each party, which might include the commitment of the client to source the part or subassembly to the supplier in charge of the design. However, it is desirable to limit the ESI contract to the design phase, and not merge it with the sourcing arrangement, in order to avoid excessive complexity.

The example of Philips Floor Care looking around for a new supplier is quite atypical. It underscores the problems inherent in not having built enough trust over time with the company's supply base. Because they were unable to cultivate positive relationships with their suppliers, Philips found themselves in the difficult situation of having to build trust from scratch. Trust-building, as we have seen, is a very delicate process that takes time, experience and patience. This is why, in particular, ESI should not be seen as a 'fad' which companies can adopt and give up at whim. It relies on a long-term policy that aims to develop the supply base to the mutual benefit of both parties. The companies that embark on this journey should set their course in this direction almost indefinitely.

8 The Strategic Implications of ESI

'Working on the Mustang was challenging,' according to Lee Davidson, manager of automotive products for Intel. 'The challenging part was meeting the schedule that Ford required.'[145] Intel supplied some of this automobile engine's electronics, and was one of around 200 suppliers that were swept up in Ford's move from the 48-month 'Concept to Customer' programme to its new 37-month 'World Class Timing' new car development process.

The (Lost) Year of the Mustang

In 1991 Ford sent 400 'car nuts' – managers, purchasers, engineers, manufacturers and suppliers – to an abandoned building in Dearborn, Michigan with the mission of relaunching a car that had been introduced in 1964 and untouched since 1979. The new Mustang was to have the acceleration and handling of a modern sports car, state-of-the-art electronics and sound, be priced at less than $15 000 and delivered in 37 months instead of the 48 normally required. In December 1993 this new model reached dealer showrooms: it carried a sticker price of $13 365 (without options), was sleek, powerful and had gone into production in 35 (not 37) months. Ford had invested $700 million in the new platform and projected 8350 sales in the first month; 11 201 cars were actually sold and the 'World Class Timing' process that made it all possible is now used in all of Ford's new car developments. To achieve these production targets, Ford set a number of process innovations in motion.

The first was the creation of a co-located cross-functional team. Under the 'Concept to Customer' process, different geographically dispersed and functional (engineering, purchasing, manufacturing) Ford teams passed their development work back and forth on a scheduled basis. 'The individual teams would meet on a periodic basis, but they were not located in one place,' said Steve Vince, manager of vehicle programmes at Ford's electronic division (ELD).[146] Ford decided this was unworkable if it was going to reduce the car's cycle time. A cross-functional team of its own personnel and suppliers was created at the project's outset and this group worked together continuously, to the satisfaction of many: 'It seemed that the team was made up of a bunch of car nuts,' Vince said. 'Everybody truly loved the platform and there was a willingness to do whatever it took to bring this car home.'

Doing more business with fewer suppliers became a top priority for the team, so the new Mustang's supplier base was cut from about 300 to 200. During the 1980s, Ford had evaluated its supplier base in conjunction with new production strategies. This resulted in a group of preferred suppliers of whom there were about 1800 in 1992 for North American operations. Consistent with corporate strategy, this number was reduced to 850 by 1994 and according to Norm Ehlers, vice president of purchasing at Ford, the move was a key to World Class Timing: 'I think we would have had a difficult time with World Class Timing if we hadn't reduced our supplier base.'[148]

Nor did Ford use competitive bidding with the Mustang's suppliers, and this was the first large-scale effort that disinherited the concept. Previously, Ford sent requests for quotations to 20 suppliers or more but for the new Mustang, the objective was to develop the kind of relationship described throughout this book with a small number of design-level suppliers. The Mustang's suppliers were virtually guaranteed to keep the business throughout its product life and to obtain Ford's commitment for future business. In return, Ford expected a lot from the suppliers. They had to fulfil evolving quality, delivery and innovation requirements, and work to reduce the cost of parts supplied. They also had to stay competitive: if another supplier showed Ford a better design or process capacity, that supplier could replace the current one. 'This is the new tradition,' said Vince. 'We are using it to change the culture at Ford.'[149]

The 1964 Mustang differed from its 1994 counterpart in more than its specifications. The 1964 model is a representation of technological capabilities that were a product of their time, and these exist on an evolving continuum of manufacturing know-how. While the 1964 car was a sequentially developed success in a relatively benign era of marketing-dominated manufacturing, the 1994 Mustang was conceived by a co-located cross-functional team that had considerably more design and manufacturing discretion than the functional departments that preceded them. It is also true that World Class Timing has some corporate antecedents that add historical insights. Ford's Cortina held a 12 per cent share in the British compact car market through 1982 while both Ford and GM were preparing new model introductions. GM was the first to introduce its new Sierra in 1983, a full year ahead of Ford. The Cortina's market share dropped to 8 per cent as the Sierra gained a 9.5 per cent share in 1983. To recoup this loss, Ford had to add value to its new model at a reported cost of around $1 billion in lost profit over the ensuing five years.[150] In 1988, Ford acquired Jaguar, which subsequently experienced a 1990 sales drop of 43 per cent due to tough competition from prestige Japanese cars like the Lexus and Infiniti. Jaguar's sales fell another 54 per cent in the first six months of 1991 while both the Lexus and Infiniti recorded gains.[151] Jaguar's management criticized Ford's slow, bureaucratic (five- to six-year) new product development process for being unable to meet the Lexus and Infiniti head-on; in the interim, Jaguar management made do with reworking and adding value to the existing platform.[152]

The Evolving Practice of New Product Development

A recent analysis of 203 new product development projects (NPD) in 123 industrial product firms[153] concluded that the leading cause of new product failure was a lack of attention to pre-development activities, at least some of which (JIT arrangements with suppliers, intra-firm horizontal integration, speed-based strategies, cross-functional development teams) were unknown in 1964. The 30 years separating the two generations of Mustangs have witnessed an evolution of NPD practices that is, in a word, dramatic. Today, most experts agree that 'Linkages, integration, synergy, competitive advantage, distinctive competence, and interdependence ... represent the strategic battle cry of management in the competitive environment of the 1990s.'[154] This has certainly been the case at Ford. The company considers its drive trains (engines and transmission systems) to be distinctive competencies and retains the design rights and produces them itself. The remaining components are divided into two groups: those provided by 'certified full service suppliers' (in the ESI tradition) and those manufactured by subcontractors working from Ford designs.[155]

When we consider the evolution of NPD strategies at Ford and other leading producers, ESI can be situated within the framework of cutting-edge supply chain management. CAMI, the Consortium for Advanced Manufacturing International, defines best practice supply chain management as a 'structured business approach to balance, synchronize, and synergize all internal and external resources and assets to achieve competitive success'. This association believes that, in the future, supply chains will compete with supply chains, not companies with companies.[156]

ESI has also been located in a concurrent engineering framework for new product development, an approach which proponents say delivers 'a faster product-design cycle because it anticipates problems early on and balances the needs of all parties, including those of customers, engineering, suppliers, marketing, and manufacturing'.[157] But supply chain management and concurrent engineering are not two different approaches to best practice NPD; they are complementary and overlapping concepts that are subsumed in a larger set of developments. To grasp the full implications of ESI in current NPD strategies, it may be helpful to revisit the issue of innovation itself.

The Nature of Innovation

Innovation is a phenomenon with several important facets. As commonly employed in business, it is perhaps first and foremost an analogue to the human act of creation which most Western minds connect to those creative individuals – Branson, Edison, Honda – who (successfully) challenged prevailing ideas and institutions. But the imagery of 'great men' is a psychosocial curio since, brilliant ideas notwithstanding, innovation is a fundamentally social phenomenon (as we explained in Chapter 1). Indeed, economists consider it to be the ultimate source of value creation in society and in business. Britain's Science Policy Research Unit was the first to document innovation's 'social' dimensions by identifying the importance of external technology sources in new product development. Today, virtually all of our research knowledge situates the innovation process within a network of actors.

Meanwhile, at the heart of innovation lies a chaotic process that combines creativity and risk, one that many managers describe as improvised and unsystematic. Research agrees that the likelihood of innovation is enhanced by organizational conditions of slack, loose controls, few constraints and the freedom to safely think and act in non-traditional ways. Peter Senge, for example, argues for a context that permits reflection, experimentation and contemplation in order to move beyond existing patterns. Ed Schein suggests that human systems must be unfrozen and motivated under low-threat conditions in order to 'reframe or cognitively re-define some aspect of their psychological field. This model accounts for creativity at the individual level and describes a generic change process.' In fact, both research and experience agree that slack, appropriate (loose) controls and the freedom that comes from a sense of psychological security are important to creativity and innovation.

Most of this knowledge is rooted in micro-psychology, the dynamics and determinants of individual creative action. Some of it has climbed a step above to assess the conditions leading to (or at least encouraging) organized innovation. It has been suggested, for example, that four different 'inquiring systems' characterize the way organizations approach learning and innovation: by closing themselves to the external environment and working with tradition (Leibnizian), by seeking consensus within a bounded domain (Lockean), by producing multiple feasible alternatives that expand the domain (Kantian), and by producing conflicting alternatives that demand resolution within the domain (Hegelian). This implies that an organization will have innovation propensities or impediments based on the implicit assumptions in its new product development culture. The concept of a unitary organization and its innovating culture is increasingly fluid, however: where one finds organizational innovation today, one generally finds organizational partners.

ESI Partnerships and Innovation

The number of alliances, collaborations and other forms of inter-organizational partnership has grown dramatically in the past decade. Of these, vertical integrations are perhaps the most important engines of product innovation. Why? Because the basic criteria for new product success – speed, quality and cost – are now difficult if not impossible to achieve alone. Opinions vary on the fundamental drivers of this phenomenon but as we discussed in Chapters 2 and 4, the majority agree that technological complexity and change are among the most significant. Increasingly, the response is to involve suppliers in the innovation process itself through ESI.

Rosabeth Moss Kanter (1994) claims that ESI is among the most demanding forms of partnership because it federates the core competencies of each partner around a project entity and, as we outlined in Chapter 4, the stakes can be high for both. The SMART partnership, for example, reaches deep into the heart of both the producer's and the supplier's businesses. But like others, these manufacturers are measuring benefits in proportion to such risks – indeed, some view them as vital opportunities. Most analysts now conclude that producers who master the dynamics of industrial partnership have a significant competitive advantage. A subset places the matter on an evolutionary timetable (Chapters 1 and 3) where new product development moves from internally focused and self-articulating practices to today's integration, flexibility and networking between partners. Indeed, many companies now assume that without innovation-oriented partnerships that accent time, quality and cost, they are courting catastrophe. It is in this regard that firms which embark on ESI must recognize its place in a larger strategic landscape. With internal capabilities straining under product proliferation, technological complexity and competitive pressures, supplier inputs are becoming essential elements in a network conception of how external sources of expertise can leverage innovation. This leads us to define five key implications of ESI and strategic management.

1. The End of Self-sufficiency: Innovation and Networks of Interdependency

ESI is one part of a producer's network of interdependencies, but only one part. As noted throughout this book, the pre-design and design stages of NPD are now crucial to its success, leading to the increasing importance of R&D networks, lead users, pre-competitive research, industry associations, webs of best practice knowledge, personal relationships between scientists and engineers, and systems of market intelligence. The required shift in strategic thinking is epochal: from a business firm at the centre of its self-sufficient universe, to a business universe of strategic centres to which an individual firm must connect.

James Moore recently adopted this logic to argue that 'industry' has lost its usefulness as a business concept.[158] Companies scrambling for supremacy in direct competition with one another is an idea whose time is past. Corporate boundaries are not only vague today, in many cases they are thoroughly undefinable. Moore suggests we replace 'industry' with 'ecosystem' to capture this reality: ecosystems of firms that live together in cooperation and compete within that context. The ecosystem metaphor highlights the fact that competition is changing form to co-exist with cooperation within the business context. It is now clear that companies are finding the need, and the ways, and the means to co-exist and co-evolve with one another in a proximate environment. This process involves cooperation as well as competition and can take the form of generating shared visions, forming alliances, doing pre-competitive R&D, forming ESI design-based partnerships and managing complex relationships. Moore evokes, for example, the March 1996 announcement of overlapping alliances among AT&T, America Online, Microsoft and Netscape to provide Internet services – all of whom except

AT&T have competing software products. The ecosystem metaphor thus makes a link to biological evolution and the Darwinian processes by which entities adapt and strengthen their kind. Co-evolution within a context implies that all entities are interdependent, such that cooperation is as necessary to adaptation as competition and conflict (which are intensifying). The key is understanding when to engage which process, and to what effect.

Working with direct competitors on shared research can improve conditions for everyone; strengthening lead-user and supplier relationships can similarly benefit the company. Moore cites as an example the experiences of Asea Brown Boveri (ABB), a large firm headquartered in Zurich that provides power generation systems and electrically powered industrial equipment. Paul Kefalas became CEO of ABB Canada in 1994 and faced the problem of stagnating sales. The classic remedy would have been to upgrade and push his products and processes, but Kefalas did the opposite: he began to cooperate with the company's business environment. ABB Canada asked a number of leading firms to share the details of their strategy and began working cooperatively with several regardless of whether they were ABB customers. When willing, ABB assembled a group of experts from across its operations and set them to work with the firm to develop creative ways of realizing its plans. In one case, the ABB experts worked with a large mining company to reduce its production costs while creating safer working conditions; the collaboration produced mining robots that are remotely controlled by technicians in offices. More than a dozen lead-user collaborations or joint ventures were thus established by the end of 1995 and, unsurprisingly, ABB Canada's sales strongly improved.

For strategists plotting the future of their firms, the research concerning the implications of supplier involvement is clear: ESI relationships between suppliers and producers will increase in number, degree and importance. Lean manufacturing, concurrent engineering and TQM trends will not only emphasize JIT production and delivery of components and subsystems, they will also implicate the design level of a firm's activities.[159] The synchronization thereby required calls for more cooperation between suppliers and producers than has ever existed in the past. The (vertical) integration of the supply chain within a company's business functions implies that suppliers, producers and users will cooperate and co-evolve as never before. Concurrently, supply management will evolve as a strategic strength of the firm to include notions of 'capability buying' where a supplier's design, development and process competencies are relied on for speed and innovation.[160]

2. The New Drivers and Variables are Time, Complexity and their Implications

Strategy begins with as clear an understanding of the factors underlying a firm's success as possible. In this regard and relative to ESI, we emphasize the point that a number of drivers and variables are in play. The factors pushing firms toward ESI have been noted throughout this book but perhaps pre-eminent among them is complexity, a dimension of business reality we can extract from experiences with technology, business processes and networks. In brief:

- Products are incorporating an increasing variety of technologies.
- Each technology is continually evolving toward a more sophisticated state.
- The growing mix of technologies used in a product increases production complexity.

The growing variety of technology in products and the complexity of the production process that results, in addition to the explicit strategy of product proliferation noted in Chapter 1, generates a heavy demand on organizational resources. Strategists in a number of industries are thus increasingly oriented toward outsourcing and ESI as a way of facilitating product development and concentrating on core competencies. These, in turn, are becoming best business practices which other firms emulate.

With complexity perhaps chief among the fundamental drivers of ESI, time and quality become the other key variables affecting success in the marketplace. At the beginning of this decade Honda had the ability to take a car from design to market in 2.5 years; in contrast, Ford required five years and GM eight.[161] Honda's time-to-market has been an important competitive weapon: while Honda was developing its third-generation product, Ford was working on its second and GM was still ironing out bugs in the first generation. While this is historical fact in the automotive industry, the CEO of Hewlett-Packard, John Young, claims the same for office equipment. Reduced time-to-market is making the competitive difference for HP due to the life-cycle compression in HP's product lines. In the past, some products had life cycles of two decades with the average being around seven or eight years; today, HP's computers, printers and fax machines have life cycles of 18 months or less. Young concludes that being a year late to market means a lost window of opportunity, an obsolete-on-release new product and market share forever lost to competitors.[162]

The claim that complexity and time are two fundamentals underlying the adoption and management of ESI also finds support among futurists working in related areas. MIT's Sloan School recently undertook a multi-year research initiative termed 'Inventing the Organizations of the 21st Century'[163] and, through one scenario group, discerned organizational futures by considering a range of forces, uncertainties and logics that seem to be shaping organizations. The conclusion: five future forces are likely to be the most important, in this order:

1. Technology
2. Human aspirations[164]
3. The global economic, political and physical environment
4. Complexity[165]
5. Demographics[166]

Time, technology and complexity, we would argue, are among the most persuasive factors influencing top managers to include ESI in their strategic portfolio.

3. To Innovate is to Learn: ESI, Organizational Learning and Knowledge Management

It can be said that organizations succeed based on what they know and how well they harness their knowledge. This is the banner flying above the current wave of 'organizational learning' and 'knowledge management' initiatives. From this perspective, and in the final analysis, intellectual capital is located at the genesis of all economic wealth and value. This concept is then characterized and categorized in various ways to achieve a certain practicability in contemporary business. Land, labour and financial capital, for example, become second-order constructs rooted in an earlier socio-economic paradigm: without intellect behind any of the three, nothing of real value can be produced.

Both research and experience claim that product innovation can lead to organizational learning and knowledge management. We noted in Chapter 1, for example, that Leonard-Barton defines core competencies in terms of a firm's technical, managerial, human and knowledge systems.[167] A company draws on these basic resources to develop in the face of external demands. If the firm's knowledge becomes a 'core rigidity', the outlook is bleak. The issue is therefore the evolution of core competencies and Leonard-Barton suggests that this is the central role of innovation. New product and process developments are opportunities where organizations can introduce new capabilities, question those currently relied upon, offer alternatives and, in short, learn. Nonaka and Takeuchi take a similar position, arguing that learning results from the interaction between tacit and explicit knowledge in an organization.[168] Properly managed, a four-fold set of processes accumulates knowledge from which innovation opportunities both result and arise. At a time when the basic assets of a firm are no longer its land, labour and financial capital, product innovation thus becomes a critical process that generates and solidifies organizational knowledge.

Knowledge management has both internal and external components. There are many works outlining the internal conditions that lead to knowledge generation but fewer that underscore the importance of external actors. The approach is usually cognitive with the implication that traditional processes can inhibit learning and knowledge creation: the master plan – often a result of fixed views and hard mental boundaries – spawns goals, objectives and targets that are identified and then imposed. People and units therefore adhere to the hierarchy and ignore change events in their task and contextual environments.

In fact, it is clear that like any closed system, a social entity that remains closed to new or novel interactions within its context is in a static state, and stasis in the current business environment is a prescription for failure. The process of learning hinges on the ability to be open to changes in an environment, on the ability to examine and challenge fundamental operating assumptions, and to change. The appropriate strategy is one based on variety, experimentation and flexibility in search of greater adaptation to the relevant context. This 'learning to learn' is becoming a significant, if popular, business capability. Pedler *et al.* argue, for example, that strategy should be continuously reformulated as experience accumulates, and that management should be treated as a continuous experiment rather than a fixed set of solutions.[169] To the extent a company's strategists adopt ESI as a way of boosting product development or innovation capabilities, they set a course toward organizational learning and knowledge management.

4. The Emerging Organization: Integrated and Interconnected

Strategists that operate ESI successfully in their firms have built or are building organizations with cross-functional, vertically integrated systems and structures. Cross-functional work is becoming the norm in product development organizations: according to a recent survey of engineering firms, 90 per cent of the best-in-class relied on cross-functional teams – but so did 82 per cent of those that failed to qualify as 'best-in-class'. These cross-functional teams had final authority on project deliverables, over and beyond an organization's functional managers, and in 70 per cent of the best-in-class companies (and 58 per cent of the rest) they had direct control of the budget as well.[170]

Suppliers are increasingly involved in cross-functional teams with their manufacturing clients and in some cases, their participation is required. Beyond this, the cross-functional structure can be seen as a pragmatic adaptation to the networked realities in which many companies find themselves. The traditional (Porterian) value chain, for example, is under attack for being linear, sequential and ignorant of the feedback loops that are teeming along the chain. Normann and Ramirez propose the richer notion of value constellation in its place.[171] Value constellations are also founded on value-adding activities but these are not necessarily linearly related: activities can take place simultaneously or sequentially in a value constellation, performed by different actors on an exceptional or regular basis. The value-constellating organization is, in other words, highly flexible and adaptive, and it quickly calls traditional organizational boundaries into question. Rather than having fixed boundaries between organizations, the boundaries separating organizations are recognized as flexible and fluid.

At what point, for example, and according to what logic are the supplier in-plants at the Bose facilities in Boston 'Bose employees' or those of the supplier that puts them on loan? What value are they adding, to whose product, for the ultimate benefit of whom? Questions like these are plaguing organizational theorists, as they have for a decade or more, and their influence promises to increase rather than diminish. ESI is a vertical integration that puts traditional definitions of the organization in severe question.

Rather than repeat the management mantra of downsizing, outsourcing and networking as the organizational designs of the future, we offer two ESI-relevant reports. One comes from Robert Lutz, Chrysler's president, who (as we noted in Chapter 3) said, 'What we're finding is that having fewer people and limited resources doesn't have to be a liability – not if those resources are properly focused and not if your people are organized ... .'[172] This is in line with Nokia's US personal computer display division which accounted for sales of over $150 million in 1995 with only five employees. This five-person core oversees an outsourcing network that, naturally, includes suppliers.[173] While some in the US may not be involved with the design of computer displays, others most certainly are and particularly so in Europe.

One part of the MIT scenario group mentioned earlier believes this may be a template for 21st Century organizations. Another wonders if supra-national 'virtual companies' will dominate the organizational landscape. In either scenario, there is agreement that firms will be increasingly networked to one another, and that suppliers at all phases of the value constellation will be integrally involved. 'Permanent or semi-permanent keiretsu-style partnerships between major manufacturing companies and their families of captive suppliers ...' seem to be an emerging reality, conclude the authors.[174]

5. Innovation Leverage: the Strategic Agenda

Finally, a move toward ESI implies a strategic decision to leverage innovation for the multiple benefits it provides an organization. As we have detailed throughout this book, these benefits fall into two major categories: economic and organizational. Economic benefits arising from innovation are those that relate to market/consumer acceptance on one hand, and overall profitability resulting from internal efficiencies and lowered transaction costs on the other. Organizational benefits are cast as in this chapter in terms of the efficient, integrated, networked systems and structures which appear to be an expression of 21st Century realism.

The Requirements of Innovation-Based Partnership

The mechanics of how this obtains from the adoption of ESI are perhaps clear to readers by now. Early Supplier Involvement blends the competencies and expertise of one organization with another to form, if briefly, one of the most concrete examples of the 'new', 'networked' organization in evidence today. Based on the research programme described in Chapter 4, we fix the pivot point of this innovation-based partnership in the idea of *inter*dependence. It is interesting that while companies and managers have experience with being independent and competitive, or dependent and vulnerable, very few have developed the skills for managing interdependence.

Interdependence is both the strategic interface and the operating framework for companies like Motorola, Honda and Nokia. Strategists in these companies are honing core competencies by accessing design-level expertise in suppliers, and engineers are adjusting to life in cross-functional project groups.

The Sony managers who pointed out that 75 per cent of their effort was focused, not on the company's manufacturing process, but rather on that of its suppliers, typifies an ESI relationship built on trust, openness, and cooperation – values that do not characterize business relations for many managers. Contracts with clearly defined goals, plans, forecasts, costs, benefit distribution and legal remedies are the norm. At root, such formalisms are aimed at maximizing gains, which are usually zero-sum. They should not, however, characterize best practice ESI nor the theoretical ideal for innovation-based partnerships.

Business relationships can be defined by a number of dimensions; the precision of legal contracts, for example, is one way of fixing responsibilities, distributing benefits and reducing uncertainty. We found that the dynamics of innovation and partnership have a more substantial foundation in psycho-social dynamics, however. Our research indicates that there are at least five dimensions that define an effective innovation-based partnership. These relate to power, structures, formality, communication and benefits as displayed in Figure 8.1.

Our proposition is that choices toward the right of these continua will facilitate partnership-based innovation more than choices toward the left. We arrive at this through field research: best practice innovators appear to be operating toward the right. But this is also supported theoretically given the nature of contemporary innovation, the requisite psychosocial

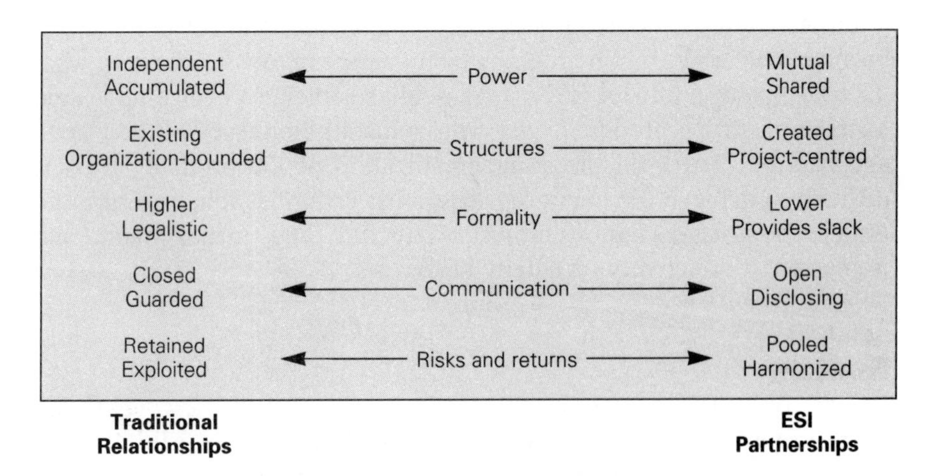

Fig. 8.1 The five dimensions of innovation-oriented relationships

dynamics, the evolution in organizational forms and functions, and the high-stakes uncertainty that accompanies most new product development efforts. A brief discussion of each dimension will clarify matters.

Power

ESI redistributes power between clients and suppliers. Traditional clients, especially their purchasing departments, often have difficulty relinquishing control to the supplier: they insist on fixing specifications as tightly as possible in order to minimize risk and uncertainty. ESI, on the other hand, acknowledges the need for mutual influence because it is difficult (if not impossible) to pool the competencies of multiple actors effectively when one or more is unable to invest in, change and influence the larger process. This also mirrors the need for loose control and freedom inside the organization to promote creativity and innovation. Such conditions are difficult to achieve across organizations when a manufacturer seeks to replicate the control it enjoys internally with external parties.

Structures

Beyond the cross-functional project team, ESI reaches toward a cross-organizational model of management. Blending the work of two companies on this scale requires that structures, systems and methods are all brought into a high degree of alignment. The anchor, the immutable point of reference, is the project itself which, from another perspective, becomes the organization around which the rest revolves. To recap what a Kodak manager told us, ESI's two essential elements are 'an empowered team and a strong product leader who holds all the pieces of the puzzle and has the matrix on the commercialization of the product'. This is difficult to achieve for many organizations: it implies that two or more organizations agree on a project's structure and management, and implement it effectively on a daily basis.

Formality

Many first-time ESI efforts stumble on four issues: intellectual investments, expected outputs, confidentiality and pricing. These are often at the centre of concern for ESI partners and nearly all the companies in our research said they presented significant challenges. But interestingly enough, only two of the 25 ESI organizations we studied could point to concrete problems in this regard. Indeed, those with generations of ESI experience displayed a remarkable degree of interorganizational informality: some had jettisoned contracts to operate with handshakes and personal relationships, others had only loosely defined agreements with their suppliers. This level of interorganizational slack is made possible by long time horizons and trusting relationships.

Communication

The communication flows in a multi-party innovation project must be excellent within and between the organizational units involved. Effective ESI relationships are based on effective information flows across company boundaries. Best practice in this area reinforces cross-functional work with sophisticated electronic communication technologies that Roy Rockwell, a leading researcher in the area, has termed nothing less than 'dramatic' and 'radical'. These include groupware, data systems, expert and simulation modelling systems, and CAD and CAE systems that are compatible across both organizations and units. The goal is that nothing impedes the quick and accurate transfer of innovation knowledge between firms, functions or individuals.

Risks and Returns

ESI partners make investments with a view to future outcomes; they pool the risks and the returns on a jointly developed product. As economic agents, they can only do so given a perception of fair benefits distribution. Research shows that the perception of fairness ensures the stability of such partnerships. Fairness does not denote equality, however: expected returns need only be in proportion to the contribution of both parties. Neither are successful ESI partnerships obsessed with fastidious accounting (unless the balance is radically tilted). Instead, they focus more on growing the pie than sharing the pie. Experience with numerous ESI relationships indicates that the former attitude is more common than the latter.

Conclusion

We hope that this and the preceding discussion has made it clear that ESI will stay on the management agenda. This is not necessarily because the challenges are intractable, although most companies that implement ESI find it harder than expected. The message from one of the most experienced ESI firms, Fuji Xerox, is that they have considerable room for improvement a decade after launching the initiative. The more fundamental reason is that ESI is a 'moving target'.

The allocation of activities between a manufacturer and its suppliers is dynamic: that which is conducted internally today may be outsourced tomorrow and the converse holds as well. Indeed, activity allocation along the value chain is affected by several contradictory factors. There seems to be a trend towards externalization as products reach later stages of the life cycle. On the other hand, new technology and market-pull innovation develop product concepts for which there is no available market supply. Later, when the technology and/or product is mature and standards are established, externalization might resume. As manufacturers ride the wave of successive technologies and products, they will find themselves constantly oscillating between outsourcing and integration.

ESI constitutes a central point in this oscillation. It succeeds the stage when the manufacturer subcontracts a given activity previously conducted internally, but still according to its own technology. And it precedes the stage when the supplier becomes fully responsible for the production and the design, independent of the manufacturer's technology. Because this oscillation is constant, ESI will remain a central concern.

As well, there is little doubt that the concept of ESI itself will evolve. Effectiveness will increase as we learn from practice and are able to codify knowledge. At present, the knowledge base is developing but is still a précis of the larger domain. Since we began this research in 1993, however, work has advanced considerably and is continuing. The need to deepen and accelerate this research is evident. There are certain to be many contributions in this regard given the attention ESI has garnered over the past five years.

Notes

1. Standard & Poor's Marketscope research group, October 1996
2. Ranganath Nayak and Jean Philippe Deschamps, *Product Juggernauts*, Harvard Business School Press, Boston, 1995
3. James M. Utterback, *Mastering the Dynamics of Innovation: How Companies Can Seize Opportunities in the Face of Technological Change*, Harvard Business School Press, Boston, Massachusetts, 1994
4. W. Cham Kim and Renée Mauborgne, 'Value Innovation: The Strategic Logic of High Growth', Harvard Business Review, January–February 1997
5. Dorothy Leonard-Barton, 'Core capabilities and core rigidities: A paradox in managing new product development', *Strategic Management Journal*, Vol. 13, 113–125, 1992
6. Dorothy Leonard-Barton, *Wellsprings of Knowledge: Building and Sustaining the Sources of Innovation*, Harvard Business School Press, Boston, Massachusetts, 1995
7. Ikujiro Nonaka and Hirotaka Takeuchi, *The Knowledge-Creating Company: How Japanese Companies Create the Dynamics of Innovation*, Oxford University Press, 1995
8. Ian Barclay, 'The new product development process: past evidence and future practical application', *R&D Management*, 22, 3, 1992
9. References from Ian Barclay's review
10. Shona L. Brown and Kathleen Eisenhardt, 'Product development: past research, present findings, and future directions', *Academy of Management Review*, vol. 20, No. 2, pp. 343–378, 1995
11. Barclay, op. cit., p. 261
12. Shona L. Brown and Kathleen Eisenhardt, op. cit. p. 354
13. Eric Von Hippel, *The Sources of Innovation*, Oxford University Press, New York, NY, 1988
14. Ken-ichi Imai, Ikujiro Nonaka and Hirotaka Takeuchi, 'Managing the New Product Development Process: How Japanese Companies Learn and Unlearn', in Kim B. Clark, Robert H. Hayes and Christopher Lorenz, eds., *The Uneasy Alliance*, Harvard Business School Press, Boston, 1985
15. Kim B. Clark and Takahiro Fujimoto, *Product Development Performance*, Harvard Business School Press, Boston, 1991
16. Shona L. Brown and Kathleen Eisenhardt, op. cit. p. 364.
17. Everett Rogers, *The Diffusion of Innovation*, The Free Press, New York, NY, 1962
18. John A. Czepiel, 'Word-of-Mouth' Processes in the Diffusion of a Major Technological Innovation', *Journal of Marketing Research*, Vol. XI, May 1974, pp. 172–180.
19. Bruno Latour, *Aramis ou l'amour des techniques*, Editions La Découverte, Paris, 1992
20. Håkan Håkansson, *Industrial Technological Development: A Network Approach*, Croom Helm, London, 987
21. Francis Bidault, *Salomon: The Monocoque Ski*, IMD case study, GM 575, 1994
22. Francis Bidault, *Philips Floor Care: The Triathlon Project*, IMD case study, GM 565, 1994.
23. Fumio Kodama, *Emerging Patterns of Innovation: Sources of Japan's Technological Edge*, Harvard Business School Press, Boston, 1991. Fumio Kodama, 'Technology Fusion and the New R&D', *Harvard Business Review*, July–August 1992.
24. Kodama, art. cit., p. 71

25. Kodama, art. cit., p. 77
26. *Business Week International*, April 1, 1996, pp. 28–30
27. James Richardson, 'Parallel sourcing and supplier performance in the Japanese automobile industry', *Strategic Management Journal*, vol. 14, 1993, pp. 339–350
28. Jeffrey H. Dyer, 'Dedicated Assets: Japan's Manufacturing Edge', Harvard Business Review, November–December 1994, pp. 174–178
29. Jeffrey H. Dyer, art. cit., p.177
30. Patricia Moody, *Breakthrough Partnering: Creating a Collective Entreprise Advantage*, Omneo, Essex Junction, VT, 1993, p. 103.
31. Jeffrey H. Dyer, Dong Sung Cho and Wujin Chu, 'Strategic Supplier Segmentation: A Model for Managing Suppliers in the 21st Century', Paper presented at the 1996 SMS conference in Phoenix, Arizona, November 9–14
32. Richard Lamming (1993), *Beyond Partnership: Strategies for Innovation and Lean Supply*, Prentice Hall, Hemel Hempstead
33. Except where otherwise noted, the information on TRW presented in this chapter is based on our interview with D. Nelson, 24 June, 1994.
34. US Industrial Outlook, 1994
35. Lamming, 1993.
36. Reuters Business Briefing (Hoover's Company Profiles, March, 1997).
37. TRW's supplier relationships have recently begun to change towards a more participative style. An example of this change is the strategic alliance TRW announced with Magna International Inc. in December 1996. The aim of the alliance is (Reuters Business Briefing (PR Newswire, 16 December, 1996)): *The design, development and production of automotive products for the global market. Under the alliance, TRW will lead development efforts in occupant restraint systems, including air bags, seat belts, inflators, sensors and steering wheels, while Magna will focus on complete vehicle interiors and total body systems. As part of this strategic alliance, TRW and Magna will form and operate a technical center that will focus on total vehicle safety system integration and will support both companies in the development of systems and components.*
38. Halberstam, 1986, p. 67
39. Womack *et al*, 1991
40. The tender process outlined here can also be used for repeat contracts of existing parts, although it is also possible for the current supplier of a part to have the purchasing contract renewed automatically (i.e. without a tender process).
41. If one wins the initial business, then, assuming: (1) the manufacturer's product is successful in the marketplace; and (2) the supplier meets the terms of the purchasing contract, repeat business is more likely for that initial supplier.
42. Renewal clauses excluded
43. The preceding statement oversimplifies the situation. The tools might break down, or key employees could quit, and so on during the fulfilment of a contract with the situation needing to be rectified as quickly as possible
44. Except where otherwise noted, the information on Diebold presented in this chapter is based on our company visit (23 June, 1994).
45. Reuters Business Briefing (Hoover's Company Profiles (May, 1997))
46. An interesting footnote to this early history is that all 878 Diebold safes in the city of Chicago at the time of the 1871 Great Fire survived.
47. Reuters Business Briefing (Hoover's Company Profiles (May, 1997))
48. CAD transmission and other technological changes have also speeded things up (although these changes have also taken place in conjunction with qualified STS suppliers).
48. CAD transmission and other technological changes have also speed things up (although

these changes have also taken place in conjunction with qualified STS suppliers).
49. All examples in this and the following section were gathered during company visits during 1994 and 1995.
50. Except where otherwise noted, this section is a consolidation and summary of Womack *et al.*, 1991; Lamming, 1993; Fujimoto, 1994; and Fujimoto and Clark, 1991.
51. Quoted in Lamming, p. 17
52. Several excellent historical accounts which emphasize the contrast between the two paths of industrial development have been written about this period. Interested readers are recommended to refer to original sources (e.g. Womack *et al.* and Lamming) for more information.
53. Friedman, 1988
54. Horsley and Buckley, 1990
55. Friedman, 1988, p. 2
56. As related by Womack *et al.* (p. 48), Toyoda means 'abundant rice field' in Japanese. Thinking ahead to the marketing implications of the family name if it became a brand name, a public contest was held to select a new name for the new company. Out of the 27 000 suggestions received, Toyota, which has no meaning in Japanese, was selected as the winner.
57. Womack *et al.*, 1991, p. 63
58. Fujimoto, 1994, p. 17
59. As quoted in Fujimoto, 1994, p. 14
60. According to Fujimoto (1994, pp. 15–16), some black-box-type work had probably taken place in the pre-war locomotive and aircraft industries, but the post-war automotive industry was the first to make use of it on a systematic and widespread basis.
61. Black-box parts were outlined in Chapter 2. A black-box part is one where the development process has always been entirely under the responsibility of the supplier.
62. Except where otherwise noted, the information on Honda presented in this chapter is based on our company visit (24 June, 1994).
63. Fujimoto, 1994, p. 22
64. The term Toyota originally used for Black Box practices (Fujimoto)
65. Value Analysis/Value Engineering
66. Womack argues that human resources are the last frontier for lean production. Honda, for example, would need to appoint a non-Japanese director to the board in order to be considered a truly global lean production organization
67. The company eventually wants to be self-sufficient in Product Engineering in three geographic regions: Japan, North America and Europe.
68. As quoted in Lamming, p. 137
69. Reuters Business Briefing, 10 December 1992
70. Reuters Business Briefing, 17 February 1994
71. Reuters Business Briefing, 17 February 1994
72. Reuters Business Briefing, 25 February 1992
73. Or CSI (Continuous Supplier Involvement) as it is called at Xerox.
74. Other non-automotive examples are outlined briefly in Chapter 2. The Lexmark example is detailed in Chapter 4.
75. This section is based on interview/notes with the company (22 June 1994) and on Reuters Business Briefing (21 November 1996).
76. (Largely) excepted from Bidault and Kono
77. Subassembly with a large number of components
78. Excerpted from Bidault, F. and Butler, C. 1995. *Lexmark, A / B / C and Teaching Notes.* IMD: International Institute for Management Development. Lausanne, Switzerland

79. Nishigushi, 1994, as an example
80. The Harvard Auto Industry Project, and the International Motor Vehicle Program or IMVP. Womack, Jones & Roos, 1990
81. Nishigushi, 1994; Lamming, 1993; Clark and Fujimoto, 1991
82. Nishigushi, 1994; Lamming, 1993; Clark and Fujimoto, 1991
83. Clark and Fujimoto, 1991
84. George Homans (1950). *The Human Group*. New York: Harcourt Brace Jovanovich. p. 123.
85. Lamming, 1993; Sako, 1992
86. Nishigushi, 1994, p. 136
87. Pierre-Yves Gomez, 1994
88. Lamming, 1993
89. Womack *et al.*, 1990; Clark and Fujimoto, 1991; Nishigushi, 1994; Fujimoto, 1994
90. See, for instance, Håkansson, 1987; Callon, 1989; DeBresson and Amesse, 1991
91. Fujimoto, 1994
92. Kamath and Liker, 1994
93. See the literature review by Ian Barclay, 1992; Von Hippel, 1988
94. Håkansson, 1987; Callon, 1989; DeBresson and Amesse, 1991; Laage-Hellmann, 1987
95. Clark and Fujimoto, 1991; Womack, Jones and Roos, 1990
96. Brown, Shona and Eisenhardt, Kathleen. Product development: Past research, present findings, and future directions. *Academy of Management Review*, 20(2), 1995: 343–378.
97. Doughty, D. Understanding new products for new markets. *Strategic Management Journal*, 11, 1990: 59–78. Doughty, D. Interpretive barriers to successful product innovation in large firms. *Organization Science*, 3, 1993: 179–202.
98. Ancona, D. & Caldwell, D. Beyond boundary spanning: Managing external dependence in product development teams. *Journal of High Technology Management Research*, 1, 1990: 119–135. Ancona, D. & Caldwell, D. Bridging the boundary: External process and performance in organizational teams. *Administrative Science Quarterly*, 37, 1992: 634–665.
99. Imai, K., Ikujiro, N. and Takeuchi, H. 1985
100. Brown, Shona and Eisenhardt, Kathleen. Product development: Past research, present findings, and future directions. *Academy of Management Review*, 20(2), 1995: 371.
101. The original formulation is from E.H. Schein, *Process Consultation*. This list was adapted from Gordon, Judith (1991) *Organizational Behavior*, Boston: Allyn & Bacon, 212.
102. Adapted from Gibb, J. Defensive communication. *ETC: A Review of General Semantics*, 22, 1965.
103. Doughty, D. Understanding new products for new markets. *Strategic Management Journal*, 11, 1990: 59–78.
104. Ancona, D. and Caldwell, D. Bridging the boundary: External process and performance in organizational teams. *Administrative Science Quarterly*, 37, 1992: 634–665.
105. Seitz Corporation, Torrington Industrial Park, P.O. Box 1398, Torrington, CT. This information was taken from the Seitz homepage at http://www.seitzcorp.com
106. Plastech, 20 North Lake Street #210, Forest Lake, MN 55025. This information was taken from the company's homepage at http://www.industry.net/c/mn/03tvlcp001/03tvl
107. Asmus, David and John Griffin. AHarnessing the power of your suppliers.' *The McKinsey Quarterly*, No. 3, 1993, p. 65.
108. We are using this term in the sense defined by Jon Ricker, Manager of purchasing operations for Case Corporation in Racine, Wisconsin: 'Strategic sourcing is a systematic process that directs purchasing and supply managers to plan, manage, and develop the supply base in line with a firm's strategic objectives. Strategic sourcing, seen another way,

is the application of current best practices to achieve the full potential of integrating suppliers into the long-term business process.' *National Association of Purchasing Management (NAPM) News*, May 1997.

109. 'The Synergy of Strategic Sourcing.' An electronic article found on the *National Association of Purchasing Management* Web site, published 1997
110. This is in considerable contrast, for example, with the following statement: '... the supplier being an independent, or at least semi-independent, enterprise must have a strategy to be competitive in the industry and thus generate profits and give opportunities in the labor market' (Grant and Gadde, 1984, p. 65).
111. Sharon LeGault, Marketing Manager at Seitz. From our interview at Torrington on 30 June 1994.
112. Ricardella Valente, Carello (MM). From our interview at Torino on 1 September 1994.
113. Joseph Rizzo, Director of Program Management at Nypro. From our interview at Clinton on 30 June 1994.
114. From *The Purchasing Futures Research Project*, sponsored by the Center for Advanced Purchasing Studies (CAPS). Reported in *Clarifying a Strategic Vision for Managing Your Suppliers, Supplier Selection & Management Report*, April 1996.
115. Bruce Grant, Director of Arthur D. Little AB. Sweden. From our interview at Göteborg on 26 April 26 1994.
116. From *The Purchasing Futures Research Project*, sponsored by the Center for Advanced Purchasing Studies (CAPS). Reported in *Clarifying a Strategic Vision for Managing Your Suppliers, Supplier Selection & Management Report*, April 1996.
117. Ricardella Valente, Carello (MM). From our interview at Torino on 1 September 1994.
118. Edouard Pfister, Sonceboz. From our interview at Sonceboz on 24 August 1994.
119. Asmus, David and John Griffin. 'Harnessing the power of your suppliers.' *The McKinsey Quarterly*, No. 3, 1993, p. 68.
120. Francis Bidault, *Philips Floor Care: The Triathlon Project*, IMD case study, GM 565, 1994.
121. Oliver Williamson, Opportunism and Its Critics, *Managerial and Decision Economics*, 1993, vol. 14, pp. 97–107
122. Francis Fukuyama, *Trust: The Social Virtues and the Creation of Prosperity*, The Free Press, New York, NY, 1995
123. Alain Peyrefitte, *La Société de Confiance*, Editions Odile Jacob, Paris, 1996
124. Jeffrey L. Bradach and Robert G. Eccles, *Price, Authority and Trust: From Ideal Types to Plural Forms*, Annual Review of Sociology, 1989, pp. 97–118
125. Kenneth Arrow, *The Limits of Organization*, W.W. Norton & Company Inc., New York, NY, 1974
126. Oliver Williamson, *The Economic Institutions of Capitalism*, Free Press, New York, NY, 1985
127. Jeffrey L. Bradach and Robert G. Eccles, art. cit.
128. Peter Smith Ring and Andrew H. Van de Ven, Structuring Cooperative Relationships Between Organizations, *Strategic Management Journal*, vol. 13, 1992, pp. 483–498
129. *Business Week*, 8 August 8 1994
130. Francis Bidault and J. Carlos Jarillo, Trust in Economic Transactions in Francis Bidault, Pierre-Yves Gomez and Gilles Marion (editors), *Trust: Firm and Society*, Macmillan Business, 1997. This section will draw on this article to discuss the other dimensions of trust.
131. Mari Sako, *Prices, Quality and Trust: Inter-firm Relations in Britain and Japan*, Cambridge University Press, Cambridge, UK, 1992
132. Charles Handy, Trust and the Virtual Organization, *Harvard Business Review*, May–June 1995, pp. 40–50

133. Francis Bidault and J. Carlos Jarillo, art. cit. p. 85
134. Kenneth Arrow, op. cit., p.23
135. Francis Bidault and J. Carlos Jarillo, art. cit. p. 86. This section will build upon the discussion of the origins of trust found in this article.
136. Charles Handy, art. cit., p. 46
137. Ranjay Gulati, Does Familiarity Breed Trust? The Implications of Repeated Ties For Contractual Choice in Alliances, *Academy of Management Journal*, vol. 38, No. 1, 1995, pp. 85–112
138. Lynn G. Zucker, Production of Trust: Institutional Sources of Economic Structure, 1840-1920, *Research in Organizational Behavior*, vol. 8, 1986, pp. 53–111
139. Etsuo Yoneyama, The Importance of Mutual Trust in Japanese Business Relationships in Francis Bidault, Pierre-Yves Gomez and Gilles Marion (editors), op. cit., p.153
140. Rosabeth Moss-Kanter, …, *Harvard Business Review*, month-month, year, pp. xxx-yyy
141. Morton Deutsch, Cooperation and Trust: Some Theoretical Notes in Nebraska Symposium on Motivation, edited by W. Edgar Vinack *et al.*, University of Nebraska Press, 1962
142. Peter J. Buckley and Mark Casson, A Theory of Cooperation in International Business in *Cooperative Strategies in International Business*, edited by Farok J. Contractor and Peter Lorange, Lexington Books, Lexington, Massachussets, 1989.
143. Jeffrey H. Dyer, Dedicated Assets: Japans Manufacturing Edge, *Harvard Business Review*, November-December 1994, pp. 174–178
144. Jeffrey H. Dyer, art. cit., p. 175
145. Carbone, James. 'Ford comes up with a better way to design (reduces time to market by eliminating competitive bidding and choosing suppliers before designs are made).' *Electronic Business Buyer*, Vol 20, No 4, April 1994, p. 120(4).
146. Ibid.
147. Ibid.
148. Ibid.
149. Ibid.
150. Nayak, P. *Managing Rapid Technological Development*, Cambridge, MA: Arthur D. Little, Inc., 1991.
151. Tumulty, B., 'Luxury Car Sales', *Gannett Newspapers*, 22 September 1991
152. Memmott, M., 'Jaguar Cars Sputtering in Red Ink', *U.S.A. Today*, 9 August 1991
153. Cooper, Robert G. New Products: What Distinguishes the Winners? *Research/ Technology Management*, November-December 1990. pp. 27–31.
154. Birou, Laura; Fawcett, Stanley E; Magnan, Gregory M. 'Integrating product life cycle and purchasing strategies.' *International Journal of Purchasing & Materials* Management, Vol. 33 No. 1, Winter 1997, pp: 23–31.
155. Wines, Leslie. 'High order strategy for manufacturing.' *Journal of Business Strategy*, Vol. 17 No. 4, July/August 1996, pp: 32–33.
156. Ibid.
157. Litsikas, Mary. 'Break old boundaries with concurrent engineering.' *Quality*, Vol. 36 No. 4, April 1997, pp: 54–56.
158. Moore, James. *The Death of Competition: Leadership & Strategy in the Age of Business Ecosystems*. New York: Harper-Collins, 1996
159. Bakos, J. Yannis & Brynjolfsson, Erik. 'Information Technology, Incentives and the Optimal Number of Suppliers.' *Journal of Management Information Systems*, Fall, 1993.
160. 'Clarifying a Strategic Vision for Managing Your Suppliers.' A research study from The Purchasing Futures Research Project sponsored by the Center for Advanced Purchasing Studies (CAPS), and reported in *Supplier Selection & Management Report*, April 1996.
161. Pascale, R., 'Strategy', *Business Month*, October 1990, pp. 38–42.

162. Inglesby, T., 'Executive Interview', *Manufacturing Systems*, June 1991, pp. 2–3.
163. Laubacher, Robert J., Thomas W. Malone and the MIT Scenario Working Group. 'Two Scenarios for 21st Century Organizations: Shifting Networks of Small Firms or All-Encompassing 'Virtual Countries'?' Sloan School of Management, Massachusetts Institute of Technology. MIT Initiative on Inventing the Organizations of the 21st Century. Working Paper 21C WP #001
164. What will people in the future want? More material goods, or will they devote a greater proportion of their energies to the pursuit of non-material ends?
165. Will complexity increase, or will the limits of human capacities lead to a reaction against complexity?
166. The changes in the centers of world population gravity, and wealth shifting away from North America and Europe
167. Dorothy Leonard-Barton, 'Core capabilities and core rigidities: A paradox in managing new product development', *Strategic Management Journal*, Vol. 13, 113–125, 1992
168. Ikujiro Nonaka and Hirotaka Takeuchi, *The Knowledge-Creating Company: How Japanese Companies Create the Dynamics of Innovation*, Oxford University Press, 1995
169. An example is the recent book: Pedler, M., Burgoyne, J. and Boydell, T. *The Learning Company*, McGraw-Hill, Maidenhead and New York, NY, 1991.
170. 'New Best Practices Study Shatters Product Development Myths in Engineering Departments.' A survey sponsored by the Product Development Roundtable and reported in *Engineering Department Management & Administration Report*, June, 1996.
171. Ashkenas, R., Ulrich, D., Jick, T., and Kerr, S. *The Boundaryless Organization*. San Francisco: Jossey-Bass, 1995.
172. (Reuters Business Briefing, 10 December 1992)
173. Laubacher, Robert J., Thomas W. Malone and the MIT Scenario Working Group. 'Two Scenarios for 21st Century Organizations: Shifting Networks of Small Firms or All-Encompassing 'Virtual Countries'?' Sloan School of Management, Massachusetts Institute of Technology . MIT Initiative on Inventing the Organizations of the 21st Century

References

Ancona, D. and Caldwell, D. (1990) Beyond boundary spanning: Managing external dependence in product development teams. *Journal of High Technology Management Research,* Vol. 1, 119–135.

Ancona, D. and Caldwell, D. (1992) Bridging the boundary: External process and performance in organizational teams. *Administrative Science Quarterly*, Vol. 37, 634–665.

Antonelli, C. (ed.) (1988) *New Information Technology and Industrial Change: The Italian Case*. Dordrecht: Kluwer Academic Publishers.

Bakos, J. Yannis and Brynjolfsson, Erik (1993) Information Technology, Incentives and the Optimal Number of Suppliers. *Journal of Management Information Systems*, Fall.

Bakos, J. Yannis and Brynjolfsson, Erik (1992–3). From Vendors to Partners: Information Technology and Incomplete Contracts in Buyer-Supplier Relationships. *Center for Coordination Science Technical Report*. MIT Sloan School of Management, Cambridge, Massachusetts. First draft: May 1992; current draft: June, 1993.

Barclay, Ian (1992) The new product development process: past evidence and future practical application. *R&D Management*, Vol. 22, No. 3, 255–263.

Bidault, F. (1994) *Philips Floor Care: The Triathlon Project* (Teaching Case GM 565). Lausanne: IMD.

Bidault, F. (1994) *Salomon: The Monocoque Ski*. IMD case study, GM 575.

Bidault, F. and Butler, C. (1995). *Buyer-Supplier Cooperation for Effective Innovation*. M2000 Executive Report, IMD, Lausanne, Switzerland.

Brown, R. H. and Garten, J. E. (1994) *U.S. Industrial Outlook 1994*. US Department of Commerce, International Trade Administration.

Bidault, F and Kono, H. (1996) *Early Supplier Involvement at Fuji-Xerox, Ebina Plant*. Sophia Antipolis: Theseus and Yokohama: Keio Business School.

Brown, R. H. and Garten, J. E. (1995) *U.S. Global Trade Outlook 1995–2000: Toward the 21st Century*. US Department of Commerce, International Trade Administration.

Brown, Shona L. and Eisenhardt, Kathleen (1995) Product development: past research, present findings, and future directions. *Academy of Management Review*, Vol. 20, No. 2, 343–378.

Business Week International, 1 April 1996, 28–30

Butler, C. (1994) M2000 Interview Notes. IMD, Lausanne, Switzerland.

Callon, M. (ed.) (1989) *La science et ses réseaux*. Paris: Editions La Découverte.

Carter, C.F. and Williams, B.R. (1957) *Industry and Technical Progress*. London: Oxford University Press.

Clark, K.B. and Fujimoto, T. (1991) *Product Development Performance: Strategy, Organization, and Management in the World Auto Industry*. Boston: Harvard Business School Press.

Cole, R.E. and Yakushiji, T. (1984) *The American and Japanese Auto Industries in transition: Report of the Joint U.S.-Japan Automotive Study.* Ann Arbor: Center for Japanese Studies (The University of Michigan) and Tokyo: Technova Inc.

Cooper, R.G. (1979) Identifying industrial new product success: project new prod. *Industrial Marketing Management*, Vol. 8, 124–135.

Cusumano, M.A. and Takeishi, A. (1991) Supplier Relations and Management: A Survey of

Japanese, Japanese-Transplant, and U.S. Auto Plants. *Strategic Management Journal*, Vol. 12, 563–588.
Czepiel, John A. (1974) Word-of-Mouth Processes in the Diffusion of a Major Technological Innovation. *Journal of Marketing Research*, Vol. XI, May, 172–180.
DeBresson, C. and Amesse, F. (1991) Networks of innovators: A review and introduction to the issue. *Research Policy*, Vol. 20, 363–379.
Doughty, D. (1990) Understanding new products for new markets. *Strategic Management Journal*, Vol. 11, 59–78.
Doughty, D. (1993) Interpretive barriers to successful product innovation in large firms. *Organization Science*, Vol. 3, 179–202.
Dyer, Jeffrey H. (1994) Dedicated Assets: Japan's Manufacturing Edge. *Harvard Business Review*, November–December, 174–178
Dyer, Jeffrey H., Dong Sung Cho and Wujin Chu (1996) Strategic Supplier Segmentation: A Model for Managing Suppliers in the 21st Century. Paper presented at the 1996 SMS conference in Phoenix, Arizona, 9–14 November.
Emshwiller, J.R. (1991) Suppliers Struggle to Improve Quality as Big Firms Slash Their Vendor Rolls. *Wall Street Journal*, 8 August, B1.
Friedman, D. (1988) *The Misunderstood Miracle: Industrial Development and Political Change in Japan.* Ithaca and London: Cornell University Press.
Fujimoto, T. (1994) *The Origin and Evolution of the "Black Box parts" Practice in the Japanese Auto Industry.* Discussion Paper Series, Research Institute for the Japanese Economy, Faculty of Economics, University of Tokyo. Abridged in Shiomi, H. and Wada, K., (eds) (1996) *Fordism Transformed: The Development of Production Methods in the Automotive Industry.* Oxford University Press.
Gibb, J. (1965) Defensive communication. *ETC: A Review of General Semantics*, 22.
Gomez, P.-Y. (1994) *Qualité et Théorie des Conventions*. Paris: Economica.
Håkansson, Håkan (1987) *Industrial Technological Development: A Network Approach*. London: Croom Helm.
Halberstam, D. (1986) *The Reckoning*. New York: Morrow.
Helper, S. (1991) How Much Has Really Changed between U.S. Automakers and Their Suppliers? *Sloan Management Review*, Summer, 15–27.
Horsley, W. and Buckley, R. (1990) *Nippon: New Superpower*. London: BBC Books.
Imai, Ken-ichi, Nonaka, Ikujiro and Takeuchi Hirotaka (1985) Managing the New Product Development Process: How Japanese Companies Learn and Unlearn, in Kim B. Clark, Robert H. Hayes and Christopher Lorenz, eds., *The Uneasy Alliance*. Boston: Harvard Business School Press.
Jarillo, J.K. (1988) On Strategic Networks. *Strategic Management Journal*, Vol. 9, No. 1, 31–41.
Johnston, R. and Lawrence, P. (1988). Beyond Vertical Integration – the Rise of the Value-Adding Partnership. *Harvard Business Review*, July–August, 94–101.
Kamath, R. and Liker, J.K. (1994) *A Second Look at Japanese Product Development*. Boston: *Harvard Business Review*, November–December.
Kanter, R.M. (1994) Collaborative Advantage. *Harvard Business Review*, July–August, 96–108.
Kim, W. Chan and Mauborgne, Renée (1997) Value Innovation: The Strategic Logic of High Growth. *Harvard Business Review*, January–February.
Kodama, Fumio (1991) *Emerging Patterns of Innovation: Sources of Japan's Technological Edge*. Harvard Business School Press, Boston.

Kodama, Fumio (1992) Technology Fusion and the New R&D. *Harvard Business Review*, July–August.

Laage-Hellman, J. (1987) Process innovation through technical cooperation. In H. Håkansson, ed., *Industrial Technological Development: A Network Approach*. London: Croom Helm.

Lamming, R. (1993) *Beyond Partnership: Strategies for Innovation and Lean Supply*. Hemel Hempstead: Prentice Hall International (UK) Limited.

Latour, Bruno (1992). *Aramis ou l'amour des techniques*. Editions La Découverte, Paris.

Leonard-Barton, Dorothy (1992). Core capabilities and core rigidities: A paradox in managing new product development. *Strategic Management Journal*, Vol. 13, 113–125.

Leonard-Barton, Dorothy (1995) *Wellsprings of Knowledge: Building and Sustaining the Sources of Innovation*. Boston: Harvard Business School Press.

Milgrom, P. and Roberts J., (1990) The Economics of Modern Manufacturing: Technology, Strategy, and Organization. *American Economic Review*, Vol. 80, No. 3.

Moody, Patricia (1993) *Breakthrough Partnering: Creating a Collective Enterprise Advantage*. Omneo, Essex Junction, VT 103.

Myres, S. and Marquis, D. (1969) Successful industrial innovations; a study of factors underlying innovation in selected firms. National Science Foundation report. No. NSF 69, 17 May, Institute of Public Administration, Washington, DC.

Nayak, Ranganath and Deschamps, Jean Philippe (1995) *Product Juggernauts*. Boston: Harvard Business School Press.

Nishigushi, T. (1994) *Strategic Industrial Sourcing: The Japanese Advantage*. New York: Oxford University Press.

Nonaka, Ikujiro and Takeuchi, Hirotaka (1995) *The Knowledge-Creating Company: How Japanese Companies Create the Dynamics of Innovation*. Oxford University Press.

Piore, M. and Sabel, C. (1984) *The Second Industrial Divide*. New York: Basic Books.

Quinn, J.B. (1985) Managing innovation: controlled chaos. *Harvard Business Review*, May–June, 73–84.

Reuters Business Briefing (1987–1997).

Richardson, James (1993) Parallel sourcing and supplier performance in the Japanese automobile industry. *Strategic Management Journal*, Vol. 14, 339–350.

Rogers, Everett (1962) *The Diffusion of Innovation*. New York: The Free Press.

Rothwell, R. (1997) The characteristics of successful innovators and technically progressive firms (with some comments on innovation research). *R&D Management*, Vol. 7, No. 3, 191–206.

Sako, M. (1992) *Prices, Quality and Trust: Inter-Firm Relations in Britain and Japan.* Cambridge: Cambridge University Press.

Souder, W.E. (1987) *Managing New Product Innovations*. Lexington, Massachusetts.

Standard and Poor's Marketscope Research Group, October 1996.

Utterback, James M. (1994) *Mastering the Dynamics of Innovation: How Companies Can Seize Opportunities in the Face of Technological Change*. Boston: *Harvard Business School Press*.

Von Hipple, E. (1986) Lead Users: a Source of Novel Product Concepts. *Management Science*, Vol. 32, No. 7, 791–805.

Von Hippel, Eric (1988). *The Sources of Innovation*. New York: Oxford University Press.

Womack, J.P., Jones, D.T. and Roos, D. (1991) *The Machine that Changed the World: The Story of Lean Production*. New York: HarperCollins.

Index